SPACIOUS
Exploring Faith and Place

Smyth & Helwys Publishing, Inc.
6316 Peake Road
Macon, Georgia 31210-3960
1-800-747-3016
©2012 by Holly Sprink
All rights reserved.
Printed in the United States of America.

The paper used in this publication meets the minimum requirements of
American National Standard for Information Sciences—
Permanence of Paper for Printed Library Materials.
ANSI Z39.48–1984. (alk. paper)

Library of Congress Cataloging-in-Publication Data

Sprink, Holly.
Spacious : exploring faith and place / by Holly Sprink.
p. cm.
ISBN 978-1-57312-649-6 (alk. paper)
1. Christianity and geography. 2. Place (Philosophy) 3. Sacred space. I. Title.
BR115.G45S67 2013
263'.042--dc23

2012038416

Disclaimer of Liability: With respect to statements of opinion or fact available in this work of nonfiction, Smyth & Helwys Publishing Inc. nor any of its employees, makes any warranty, express or implied, or assumes any legal liability or responsibility for the accuracy or completeness of any information disclosed, or represents that its use would not infringe privately-owned rights.

HOLLY SPRINK

SPACIOUS

Exploring Faith and Place

Also by Holly Sprink

Faith Postures: Cultivating Christian Mindfulness

We are not angels; we inhabit space. Material—bricks and mortar, boards and nails—keeps us grounded and connected with the ordinary world in which we necessarily live out our extraordinary beliefs.

—*Eugene Peterson, Introduction to Haggai,* The Message

Acknowledgments

Virginia Woolf said that writing, especially imaginative work, is "like a spider's web, attached ever so lightly perhaps, but still attached to life at all four corners. Often the attachment is scarcely perceptible."[1] I agree. Whatever wisps of thought I may have spun here are only hanging because others have given me time and place to read, write, and explore. These people have given me real-life places to which I can attach these fragile threads. My thanks to Keith Gammons and Leslie Andres at Smyth & Helwys, Angie French, my parents, and my church family at First Baptist Church of Blue Springs. Thanks to family and friends who have read for me, let me tell stories about them, and encouraged me in writing. Thank you, Matt, Lucy, and Mikias. I love you and don't want to be *anyplace* without you.

Note

1. Virginia Woolf, *A Room of One's Own* (San Diego: Harcourt Brace, 1929) 41.

Contents

	Introduction	xi
1	The Particularity of a Place	1
2	At Home	9
3	Coming Near	17
4	Coves and Kayaks	25
5	The Late One	33
6	Here and There	39
7	The Responsive Place	47
8	Innermost Places	55
9	Potential in Locality	63
10	Panoramic View	71
11	Perpetual Departure	77
12	Transcending Particularity	85
13	Our Communal Place	93
14	Spacious Relationships	101
15	Spacious God-Country	109
16	Ripping and Sewing	115
17	From the Creative Place	121
	Conclusion	127

Introduction

And now I am here again, changed from what I was, and still changing.[1]

There is a building on the corner of 15th and Main that is a part of me. It is as much of a daily reality to me as my own hands are, and I can trace the lines of this building like the lines in my palms. I know this building. I know the entrances and exits, the least crowded bathrooms, the best places to hide if you're playing hide-and-seek.

It is not a beautiful building. Indeed, some who see it from the outside believe it to be an industrial site or manufacturing plant. The first part of the building was built in 1922. Additions were added every twenty to thirty years, each favoring its own generation's architectural trends rather than the original integrity of the building. Brown brick is next to red brick surrounded by a pothole-ridden parking lot.

At times in my life, my visits to this building were as vital to me as eating, bathing, and sleeping. There were other years when I stepped inside it as a visitor, an outsider. Now, as I return to this building almost daily, I find that I step down in places where there used to be a step or two, only to find the steps have been paved over, made safer. My body still knows this building even though my mind has forgotten.

This building is my church: First Baptist Church of Blue Springs, Missouri. I was being nourished by it even *in utero*, as my mother accompanied the hymns on the piano each Sunday. It fed me for eighteen more years until I went off to college. And thirteen years later, I returned to it alongside my husband, a minister whom the church had asked to come and begin new global missions initiatives. Now, once again, I find the rhythms of my life shaped by this building and what goes on in it. There is history here, and not just my own. Volumes of communal life have been shared here on this corner. On this corner, in this city, since 1879, God has done his right-making work.

Each time I step in the door, memories leap out at me like a game of hide-and-seek combined with *This Is Your Life*. Big life events—my baptism, my wedding—and tiny little episodes I haven't thought of in years—where my brother split his lip open, a room in which I overheard an unkind remark about myself—arrive like unannounced visitors, sometimes welcome, sometimes unwanted. I don't know whether to try to make sense of them or not. It is rich and exhausting at the same time, and leaving church some Sundays is like pushing back from the table after Thanksgiving dinner.

I remember standing outside the student building of the church before I left to go to college, telling my Sunday school teacher that I felt impressed by God to go 600 miles away to school, but I wasn't sure why. I can remember saying I was excited to see what was out there, why God might want me to go. As I moved back a few years ago with my husband, daughter, and a son on the way, I felt exactly the same. I felt certain that God had orchestrated the move, but I wondered why, after our family had expressed to him our openness and excitement to live *anywhere* in the world, he led us to Blue Springs, Missouri. We had opened up an atlas of the world to him, and the dart he had thrown for the next chapter of our lives landed just a few streets over from the house in which I was raised.

I was excited, and anxious, to move back. It was like Emily Dickinson was writing about me:

I Years had been from Home
And now, before the Door,
I dared not enter, lest a Face
I never saw before

Stare stolid into mine
And ask my Business there—
"My Business, but a Life I left
Was such remaining there?"[2]

Our family embraced the adventure and moved to Blue Springs as if we were moving to a place we'd never known. We took family drives to nowhere, just figuring out where we were now living. We read up on the Kansas City area, trying to figure out who lived here and what kinds of people were our new neighbors. We asked a lot of questions. It was a surprising adventure to see why God had led us to this neighborhood, to this corner of the world.

Now, a few years later, the adventure continues. We are still questioning, wondering who we are and how we understand God in this particular place, in our house, in our church building, in our country. The clash of my life then and my life now enables those questions. For example, if your children are eating at a chili cook-off that is held in the same place where your wedding reception was, which also happens to be the same place where you wet your sleeping bag as a child during a church lock-in (just so you know: not the same carpet), what does that do to you?

It's messy. What I mean is this: this church and this city in which I grew up were formational to my worldview, to my perception of reality and my responses to it. It is not only memories that pop out at me here in Blue Springs, Missouri; it is the culture that had a huge part in the formation of my assumptions, values, and commitments. It is what formed my patterns of thinking and shaped my faith in God. Living here is a return to what created the first lenses through which I viewed the world and the people in it. My years in global studies at seminary taught me about working through worldview change in cross-cultural settings; I never dreamed that setting would be my own hometown.

I have more questions than answers, and that's a good thing. That odd feeling, that sting that occurs when *then* and *now* collide, is a helpful prompt. Exploring where you are and why that matters to God is an incredible, ongoing process. Whether we live in the Amazon or in our own hometown, God continuously wants to widen our view of the world. He wants us to question and explore the relationship between our history, our faith, and our locality. He is pointing us to the context clues of grace all around us. He desires to teach us the purpose in our lives as we search for his kingdom perspective. And he wants us to experience his spacious, free life as we allow that purpose to affect the nuances of our cultural context.

This effort to look at our place is not a once-in-a-lifetime trip; it is a lifestyle, an ongoing conversation. As we grow in our faith and experience God's right-making ways at work in our lives, we begin to question. When we daily read Scripture, we challenge our own culture and recognize its sinful predispositions. As we encounter other places and other people, God shows us our own limitations and errant assumptions. By living in community with other believers, we see where some of our values need to be redirected. As we learn to find our contentment in God alone, we see where some of our wayward inclinations need to be cracked and shaken off. In the effort to voice the relationship between place and faith, we realize that, whether we're born in an Asian city of 25 million people or in Friendly Town, USA, worldview change isn't just a topic for missionaries or a process for people of *other* cultures. It is the normative Christian life.

Recently I read the words of the prophet Jeremiah, a person God used to relate his truth to his people. At one point in Jeremiah 24, God tells his people, "Instead of claiming to know what God says, ask questions of one another, such as 'How do we understand God in this?'"[3] I love this picture of a God who asks his people for a little less proclamation and a little more investigation. I have certainly asked questions of God before, but along this new journey, the return journey, the questions have changed. They are now not so much "What is God's will for *my* life and why am *I* here?" but "What is this place and how do I understand God in it?"

Poet and author of more than forty books, Wendell Berry also made a return journey, to his native Kentucky. "And so here," he writes, "in the place I love more than any other and where I have chosen among all other places to live my life, I am more painfully divided within myself than I could be in any other place."[4] In his essay "The Unsettling of America," he says,

> After more than thirty years I have at last arrived at the candor necessary to stand on this part of the earth that is so full of my own history and so much damaged by it, and ask: What *is* this place? What is in it? What is its nature? How should men live in it? What must I do?
>
> I have not found the answers, though I believe that in partial and fragmentary ways they have begun to come to me. But the questions are more important than their answers. In the final sense they *have* no answers. They are like the questions—they are perhaps the same questions—that were the discipline of Job. They are a part of the necessary enactment of humility, teaching a man what his importance is, what his responsibility is, and what his place is, both on the earth and in the order of things. And though the answers must always come obscurely and in fragments, the questions must be asked. They are fertile questions. In their implications and effects, they are moral and aesthetic and, in the best and fullest sense, practical. They promise a relationship to the world that is decent and preserving. They are also, both in origin and effect, religious.[5]

I'm finding that this journey of change, this daily discovery of where we are before God, is a lifelong process led by a loving Father. In voicing the relationship between place and faith, I find many paradoxes (some of which are discussed in the following chapters). As I travel this journey, I find many places where we're called to hold two opposite truths precariously in tension.

It *is* a journey about the figurative places in our lives, but it is also about the literal ones. The psalmist understood this, writing in Psalm 84:3, "Even the sparrow has found a home, and the swallow a nest for herself, where she may have her young—a place near your altar, O LORD Almighty, my King and my God." I'm finding the question I should ask is not "Who am I, God?" but "Where am I, Lord?" I'm replacing "What should I *do*, Jesus?" with "Where do you have me today?"

A narrow, dark hallway surrounds the sanctuary of First Baptist Church of Blue Springs. This hallway is curved, following the outer line of our small, rounded auditorium. The walls are made of brown brick, aligned vertically with a deep, half-inch allowance for grout between each row. While in the hallway, you can't see around the curve to the foyer; you just see hallway and the people in it.

Once, right after moving back, I was walking with a group of children through this particular hallway. The child next to me said, "Hey, do this." He opened his right hand, palm facing the wall, and dragged his fingertips along the bricks. When he did this, his fingers got tickled and scratched, going up over the bricks and down in the grout ridges. I knew this because I used to do the same thing when I was small, walking the hallway. As an adult, I followed suit that day, joining the group of children, all of whom were now running their fingers along the bricks, enjoying the texture of the place while they were there.

Do we really *know* the places in which we live? Do we recognize the feel of them? Are we acquainted with their curves? Are we on familiar terms with their people? Do we see the ways God might use us to extend his right-making love in our own neighborhoods and cultures? Are we using our senses to take in all the sights, smells, and textures? Are we letting Christ teach us his kingdom worldview and critically applying it specifically in our own contexts?

This adventure of discovery, this attempt at a theology of my place, is difficult. The benefits of this return journey, however, are sure to outweigh an unaware life soaked in certainty and nostalgia.

Join me if you like. I'm inviting you, as a child invited me, to put your hand to your terrain. As believers in Christ, we are asked not just to walk through our hallways but also to learn from the landscapes of our lives. God asks us to journey actively with him, even if we can't see to the end of the hallway. He calls us to correct where his kingdom perspective and our own views of the world don't match up. It scratches. It tickles. It is real life.

May the God who graciously gives sparrows, swallows, and us a place to nest lead us as we find our ultimate home in him alone.

Notes

1. Wendell Berry, *The Art of the Commonplace: The Agrarian Essays of Wendell Berry* (Washington, DC: Shoemaker & Hoard, 2002) 22.

2. Emily Dickinson, *Dickinson* (New York: Alfred A. Knopf, 1993) 91.

3. Passage from *The Message* translation by Eugene Peterson. Throughout the rest of the book, I will use *The Message* translation unless otherwise specified. This isn't a statement on Bible translations; it is simply the translation I was reading devotionally as I wrote this book and the one that inspired some of the illustrations and connections.

4. Berry, *The Art of the Commonplace*, 8.

5. Ibid., 22.

Chapter 1

The Particularity of a Place

Our childhood home . . . is our "first universe," and therefore becomes "the topography of our intimate being."
—Gaston Bachelard[1]

The first year I moved back to Blue Springs, life kept popping out at me in the strangest places. I would be going about my daily routine, and an emotion would come over me before I could even place the memory. Once I was walking with my daughter, Lucy, though Hobby Lobby, when an odd feeling crept up on me. I realized a few moments later that this building, which used to be the city's Walmart, was where I once got lost as a child. I remembered rushing to the service desk, where I was now paying for my Hobby Lobby purchases, and asking the sales clerk to call my mother, whom I'd wandered away from, over the intercom. The whole instance was less than five minutes long, and yet the feelings, the anxiety of being alone and the relief of reunification, crept up on me almost like a scent or a taste. If you had asked me the day before to tell you about a time when I got lost as a child, I couldn't have recalled it, and yet there it was. The emotions came over me swiftly once my body was exactly where it had been years before.

This happened frequently, and still does. Both happy and sad moments from my life come back as I find myself in parks, businesses, and schools in my hometown. It is different from nostalgia. It isn't a wistful longing but a place where time, flowing along like linear points on a piece of yarn, gets tangled up and crosses itself. Points in time in the past and future are all of a sudden cinched up together, touching each other, and you don't know what part of the yarn, past or present, to follow next.

I'm beginning to notice the connection my spirit has with the particulars of my surroundings, connections I can't cognitively enunciate. This "nudge" from my soul happens every time I drive down sections of Missouri road that have been cut through the Ozark hills, leaving exposed rock "bluffs" surrounding me on either side. It happens when I drop off my son at the same preschool, in the very same room even, that I once attended. I feel it when my daughter asks to eat at the same Wendy's restaurant where my dad took me for a Frosty on the day I flunked my driver's test. And I can't tell what is more complicated: moving to a new place and waiting for new "spirit cues" to develop over time, or returning to my hometown where my old ones are resurrected.

The answer is both. Both processes, developing connections with a new place and being willing to add new memories to an old place, are difficult, and both are also worthwhile. Both progressions prick the heart and make it swell with joy, sometimes at the same time. And I realize I would be silly *not* to acknowledge the role my surroundings play in who I've been and who I am now. "Life has supplied me with only these eyes," writes Melissa Holbrook Pierson, "only this bizarre sensibility composed solely of this accretion of embarrassingly personal, minor events that has solidified into the unshapely mass called 'me.'"[2]

Philip Sheldrake, author of *Spaces for the Sacred*, agrees. "The physical landscape is a partner, and an active rather than purely passive partner, in the conversation that creates the nature of place," he wrote. "There is an interplay between physical geographies and geographies of the mind and spirit."[3]

When my husband, Matt, and I lived in South Africa, we attended an art exhibit where South Africans had laid down on six-foot pieces of brown paper and traced the outlines of their bodies. Then, using an art therapy technique called "Body Mapping," they added details of their lives—scars, symptoms, stories, songs—to the outlines of themselves. It was fascinating to see and read the stories of so many real lives opened up to us there on the walls. We were so interested in the way that physical bodies told the story of people as well as the nation in which they lived that we decided to go home and make our own body maps. We laughed one night as we went and bought paper and crayons, outlining ourselves, adding glasses here, broken bones there. Many years and two kids later, it sounds like the kind of thing only a married couple living in another country would have time for. However, I do remember adding to my own body map the flags of all the places I had lived over the years. And it strikes me that, actually, place is kind of like that. It physically and spiritually marks you, and usually you are unaware of how it does that. How much easier would it be if we wore our places like tattoos of flags on our skin? What if each of us were able to read right on our arms or legs how the places we've lived affect who we are and who others are?

The truth is, we are able to do that. We already carry with us the effects of locality in the way we see the world and act in it. If we are honest with ourselves and take time to notice it, our relationship to where we find ourselves goes further than memories. Usually, we don't notice a place's effects on us as much as we would a tattoo on our skin, but they are just as permanent. Like contact lenses the world has put in front of our eyes, these effects thoroughly affect the way in which we view everything else. These lenses are, by definition, what we use to take in information and to make sense of situations before us. They are our worldview, our way of perceiving reality according to our assumptions, commitments, and values. As Pierson writes, "What we are is where we have been."[4]

Each day, I see new ways that locality is a huge part of my visual prescription. But even more that that, I recognize that my relationship to my place(s) is much deeper than my individual connection to it. I don't just view the world through the lens of my experience with a place; my place in the world connects me with a broader history, a communal experience, that incorporates all those who have walked, are walking, and will walk where I'm walking. "Memory embedded in place," says Philip Sheldrake, "involves more than simply any one personal story. There are the wider and deeper narrative currents in a place that gather together all those who have ever lived there. Each person effectively reshapes a place by making his or her story a thread in the meaning of the place and also has to come to terms with the many layers of story that already exist in a given location."[5]

What does this mean? It means I need to know where I live—and not just my town's latitude and longitude on the GPS. I need to know the rich history of where I'm planted, what is affecting how I grow and thrive, because, as poet Wendell Berry said, the stuff of a culture "accumulates in the community much as fertility builds in the soil."[6]

There is a difference between place and space, I'm finding out, and a lot of it has to do with our approach to a place and our intent for it. Many times, we come to a place with the idea that it is ours alone, that we can doll it up the way we want to, using it the same way we'd use a locker in junior high school. But with land, with homes, with jobs and cities, we don't just put our stuff there, clean it out at the end of the semester, and forget the combination. We don't live in a vacuum; we live in a place that has particulars about it. Where you live has a historical tone and present-day idiosyncrasies, shades of this particular people or that particular way of thinking. It is wrong to empty our places of these particularities or to travel so quickly through them that we miss them.

Wendell Berry agrees. Read his discussion of the difference between a path and a road:

> A path is little more than a habit that comes with knowledge of a place. It is a sort of ritual of familiarity. As a form, it is a form of contact with a known landscape. It is not destructive. It is the perfect adaptation, through experience and familiarity, of movement to a place; it obeys the natural contours; such obstacles as it meets it goes around. A road, on the other hand, even the most primitive road, embodies a resistance against the landscape. Its reason is not simply the necessity for movement, but haste. Its wish is to *avoid* contact with the landscape; it seeks so far as possible to go over the country, rather than through it; its aspiration, as we see clearly in the example of our modern freeways, is to be a bridge; its tendency is to translate place into space in order to traverse it with the least effort. It is destructive, seeking to remove or destroy all obstacles in its way.[7]

In our individual localities, do our lives represent paths or roads? It is our job to make sure we don't develop a resistance to the landscape, as Berry puts it, but to be aware of the ways in which our landscape informs our lives and connects them to others.

I also like how Philip Sheldrake talks about the role of place in our lives: "The hermeneutic of place progressively reveals new meanings in a kind of conversation between topography, memory and the presence of particular people at any given moment."[8] There is a conversation going on around us, and where we are living and the time in which we are living are factors—even principal players—in that conversation. We would be wise to tune in and listen rather than ignoring the implications of place on our lives. French author Gaston Bachelard, who wrote *The Poetics of Space,* goes further, suggesting that to really understand yourself and those around you, psychoanalysis isn't the key, but "topoanalysis," the exploration of identity through place, is.[9]

We must invite *faith* to our conversation table, and it must beg a seat right beside *place*. The two are old friends. For it is not only your story or my story, your city or mine, that we're really interested in. The miracle is the story of God and the way he wants to interact with us in our particular places. We each live and work in places with their own histories and with unique particularities in which God loves and works. We are not the main event; we are coming late to a meeting, to the right-making kingdom work to which God has graciously invited us.

We see in the Bible that the particular mattered quite a bit to God and his people. Details were acts of faith for Abraham, Isaac, and Jacob, even down to their earthly burial place, the cave at Machpelah. God had told Abraham that his offspring would be given the land he was standing on, a place called Shechem. Genesis 12:6-7 says, "Abraham traveled through the land as far as the site of the great tree of Moreh at Shechem. At that time the Canaanites were in the land. The LORD appeared to Abram and said, 'To your offspring I will give this land.'" Unfortunately, before that promise was realized, Abraham needed a burial place for his wife. Eventually, he would rest there, too, as well as his son Isaac, Isaac's wife, Rebekah, their son Jacob, and Jacob's wife Leah.

So Abraham purchased land from Ephron the Hittite, a transaction recorded in Genesis 23, and he buried Sarah in the cave at Machpelah. It was a seemingly ordinary transaction, except God hadn't yet given him the land. Even in the details, however, Abraham believed God's promise to him and bought land in that particular place. Those particulars were imparted to his child, Isaac, who then handed them down to Jacob, who asked his sons to take him back to the same field and cave when his time came, even though God had saved the family by moving it to Egypt during the famine.

Joseph, Isaac's son, complied, taking time to travel back to Canaan to bury his father with his once-lost-now-found brothers (Gen 50). You can imagine the impact this had on Joseph, and you can actually read what he said about his own death: "I am about to die, but God will surely come to you, and bring you up out of this land to the land that he swore to Abraham, to Isaac, and to Jacob. When God comes to you, you shall carry up my bones from here" (Gen 50:24-25, NIV).

At the age of 110, Joseph died, leaving instructions for his bones that were completed years later, when Joshua was leading the Israelites. The Bible records the event in Joshua 24:32 (NRSV): "The bones of Joseph, which the Israelites had brought up from Egypt, were buried at Shechem, in the portion of ground that Jacob had bought from the children of Hamor, the father of Shechem, for one hundred pieces of money; it became an inheritance of the descendants of Joseph." This was a seemingly minute event, and yet, if we look back to God's promise in Genesis 12, we see that Shechem was where Abraham stood and where God came to him.

This discussion doesn't mean that everyone must return to his or her birthplace. It's just one biblical example of how locality is not irrelevant to God but instead embraced in the scope of human history and his relationship to it. In faith, all of these people looked at their surroundings through God's view rather than through their limited worldview. They saw the lands they were connected to not as space but as place, filled with history, meaning, and purpose in the grand story of God's interaction with people. They saw their roles as part of that history. They saw their lives as part of God's purposes in their places. "In death," says one author, "these Israelite ancestors are no longer strangers and aliens in the land (Abraham's self-identification in v. 4), but heirs; they come to rest in the land promised them by their God."[10]

I wonder what would happen if we embraced these particulars. What would happen if we began to look at where we live with the same intentionality, the same clarity, as the biblical characters did? If we saw the role locality plays in our lives, and the role it plays in the lives of people the world over, would we approach the world differently? Would we feel a kinship and responsibility for its people? How would we care for its places?

How thankful I am that we serve a God of the particulars, one who was willing, for our sakes, to embrace a specific time and place in sending his Son in order to open up all time and all places to his grace. As Philip Sheldrake writes, "What we sometimes refer to as the 'scandal of particularity,' that God in Christ is incarnated within what is bounded and limited, is a guarantee that *every particular time* and *every particular place* is a point of access to the place of God."[11]

May we look to our place with faith and intention, with awareness and appreciation for his purposes in the world.

Discussion Questions

1. How are the places in which you've lived active rather than passive partners in your life? How have your places marked you?

2. What do you know about the history of your place? Who has lived there and who lives there now? What could you do to learn more?

3. If what we believe is true, that "*every particular time* and *every particular place* is a point of access to the place of God," what is the history of God's interaction with people in your place? Where do you see God engaging people specifically in your locality?

Notes

1. Gaston Bachelard, quoted in John Inge, *A Christian Theology of Place* (Hampshire, Great Britain: Ashgate Publishing Ltd., 2003) 17.

2. Melissa Holbrook Pierson, *The Place You Love Is Gone: Progress Hits Home* (New York: W. W. Norton & Company, 2006) 19.

3. Philip Sheldrake, *Spaces for the Sacred* (Baltimore MD: Johns Hopkins University Press, 2001) 15.

4. Pierson, *The Place You Love Is Gone*, 19.

5. Sheldrake, *Spaces for the Sacred*, 16.

6. Wendell Berry, *The Art of the Commonplace: The Agrarian Essays of Wendell Berry* (Washington DC: Shoemaker & Hoard, 2002) 189.

7. Berry, *The Art of the Commonplace*, 12.

8. Sheldrake, *Spaces for the Sacred*, 17.

9. Quoted in Inge, *A Christian Theology of Place*, 17.

10. Terence E. Fretheim, *Genesis*, The New Interpreter's Bible, vol. 1 (Nashville: Abingdon Press, 1994) 504.

11. Sheldrake, *Spaces for the Sacred*, 66, italics mine.

Chapter 2

At Home

The Christian religion is not the religion of salvation from *places, it is the religion of salvation* in *and* through *places.*[1]

I have a friend named Aisha (AH-shah).[2] She is a Somali woman in her mid-sixties who has lived in Kansas City for around five years. About fifteen years ago, she had to flee her home village in Somalia when men with guns came into her home and began to take her things. They beat her, killed her daughter in front of her, and left her and her granddaughter to die. Instead, Aisha and her granddaughter walked all the way to Kenya, finding a place to settle in the Kakuma (the Swahili word for *nowhere*) refugee camp. Eventually, Aisha was put on an airplane that landed in Kansas City. Never having had the chance to go to school, she now attends English classes every day. She grew up speaking Somali but never had the opportunity to read and write it, which makes for slow going when learning English.

Where is home for Aisha? I wonder. Though she is trying, she is not at home here. The details of her life are like that old game Boggle; they've been completely shaken and resettled into unrecognizable words that she now must try to make sense of in order to survive. Going home is not an option, and if it were, is what she knew of "home" even there anymore? She is caught, one of the world's 42 million displaced people. Though she courageously does what she can to get through each day's challenges, her new life is still unsettling. Although she has learned to recite her new address, Aisha is still homeless.

And we are the lucky ones. We live easily in our hometowns, and many of us have family nearby. We don't awaken each day wondering if we will be able to get on the right bus or if the grocery store clerk will understand us. We have all the privileges of "at-homeness," of feeling secure in our place. We have the freedom to spend time browsing the Internet or sitting in front of the television. We don't have to build relationships with the grocery store clerk, because there will be a different English-speaking one the next time we shop. We might go to a different grocery store the next time anyway because we can drive wherever we want. Everyday life is easy for us, so we settle down on the couch with a bag of chips to watch the next installment of the latest talent-search reality show. Though we are literally "at home," are we losing sight of what it means to live well in a particular place? Are we, too, displaced?

French anthropologist Marc Augé calls much of where we city-dwellers are "non-place." These are transitory places where we tend to spend more and more time, such as supermarkets, airports, or sitting in front of the computer. These places don't demand anything of us, and we like it that way. Augé says these are "curious places which are both everywhere and nowhere."[3] We are more interested in spaces than in places, it seems. We love the freedom to fashion our extra spaces, whether they be bedrooms or open evenings on our calendars, according to our individual tastes.[4]

Place is altogether different, says Augé, having three essential characteristics. Place engages our identity, it engages our relationships, and it engages our history. In other words, you aren't really anywhere unless you're willing to get involved in who you are, who you know, and your history. And, according to Augé's definition, you don't know much about others if you don't know who they are, who they know, and where they've come from.

That sounds tiring. It sounds like research. It sounds like a lot of time listening to someone telling stories of his grandma's excellent blackberry pie. (Many days, we don't want to hear those stories unless we have a slice of pie in our hands, do we?) Getting deep into our places sounds like work. It sounds like involved relationships with people. It also sounds like the people I read about in the Bible.

Over and over in Scripture, everyone everywhere has the all-encompassing need to be "at home." Whether we are Adam or Abraham, Aisha or Amy, there is something inside us that wants to feel purposefully "placed" in this world. As we read Scripture, we see the testimonies of those who have gone before with the same "itch" to be placed. Old Testament scholar Walter Brueggemann writes about this in his book *The Land*. His premise is that the whole of the Bible addresses the chief problem that all humans have: a feeling of homelessness.[5] "This sense of place," he writes, "is a primary concern of this God who refused a house and sojourned with his people (2 Sam 7:5-6) and of the crucified one who 'has nowhere to lay his head' (Luke 9:58)."[6]

If we search within and if we look at history, we find that we cannot meet this need by ourselves or in our own strength. We see in our Western culture that though we may have the glamorous privileges of freedom, mobility, and convenience, our pursuits haven't yielded that "settled" life we thought they might. Brueggemann says this failure of the urban promise, the idea that people could "lead detached, uprooted lives of endless choice and no commitment," points us toward a biblical understanding of place and faith.[7] We know what a sense of place is not, because we've lived without it. Take a look at Brueggemann's excellent differentiation between "space" and "place":

> "Space" means an arena of freedom without coercion or accountability, free of pressures and void of authority. Space may be imaged as weekend, holiday, avocation, and is characterized by a kind of neutrality or emptiness waiting to be filled by our choosing. Such a concern appeals to a desire to get out from under meaningless routine and subjection. But "place" is a very different matter. Place is space that has historical meanings, where some things have happened that are now remembered and that provide continuity and identity across generations. Place is space in which important words have been spoken that have established identity, defined vocation, and envisioned destiny. Place is space in which vows have been exchanged, promises have been made, and demands have been issued. Place is indeed a protest against the unpromising pursuit of space. It is the declaration that our humanness cannot be found in escape, detachment, absence of commitment, and undefined freedom.[8]

What is our call as Christians? Whether you live in the Amazon or in Friendly Town, USA, I believe it is to reject life as the meaningless pursuit of space and to embrace the role our places in this world play in our faith lives. It is no longer acceptable—in fact, as we read the Bible, we see it never was!—for people who follow God and his teachings to pretend we exist unaffecting and unaffected by others around us and around the globe. We must embrace the biblical implications of our desire for having a sense of place as part of our faith.

Where do we see the importance of locality in Scripture? In the beginning, we read of the swirling chaos God formed into a good place, where he could be in good relationship with humans made in his likeness. We see it in the Abraham story in Genesis 12, when God asks a man to leave his place in order that he might be a blessing to people in all places. We see the theme of land being central to the Israelites as they journey in faith out of Egypt to the land God promises to give them. We see it when Joshua is gathering territories and redefining boundaries. We see it when David is on the throne and when he desires to build a temple for God, who says, "I've never needed a house before; how can a temple contain me?" We see a longing for place in the design of Solomon's temple, in the warnings of the prophets and the yearnings of the exiled. As we digest these stories, we begin to see that "Israel never had a desire for a relation with Yahweh in a vacuum, but only in land."[9] We see that the Old Testament and the people in it wouldn't understand the idea of faith apart from ideas of place.

We see the importance of locality in the New Testament story as well. We see Jesus come and redefine what it means to be the people of God, speaking into the lives of Jewish people consumed with regaining their place. We read of a Messiah who crossed cultural boundaries one day to sit with a Samaritan woman and help her understand that true worship was not about this mountain or that holy city but about spirit and truth. We read of the incarnation, of a God who is willing to "move into our neighborhood," in John 1. We read the words of one who encountered Jesus on the Damascus road and then later wrote about our true home, "Christ in you—the hope of glory."[10] We see Paul help his own people find their place alongside Gentile "outsiders," uniting all of them. We read in Hebrews about the faith of those who journeyed but never reached their place. We see John asking the church people to open up their places and show hospitality, the mark of a true believer. We read of the culmination of time and history, a New Jerusalem, the promised place that is both "already" and "not yet." Truly, the promise of the New Testament, as well as the Old, is that of a purposeful place for those who follow in faith, the promise of "luxuriant at-homeness."[11] If we are faithful to the text and its literal and symbolic messages about place, we see that the appropriate way to view the world is through a three-fold lens that includes people, place, and God.[12]

We must understand and act on the biblical affirmation that our lives are not about space, but about place. Further, our lives, especially as Western Christians, are not about basking in our place, as though we were setting our beach chairs on the sand and looking out at the sunset. We see God's anger time and again, in the Old and New Testaments, at this privileged and presumptuous attitude. The Bible does not teach us that we have reached our places. We as Christians have not attained special favor or secret knowledge that we hold in our hands like a drink with an umbrella. We don't just get to chill on the beach until the end of history. The Bible is "not the story of land possessed but the story of land promised."[13] So how do we find the balance?

How do we work at loving and understanding the places in which we find ourselves, and at the same time not begin to believe our place is the ultimate existence?

We can learn from those who have gone before us and who teach us that we must choose the faith journey. We must choose, wherever we dwell, to be sojourners alongside each other. Brueggemann writes,

> The way of faith requires leaving a land and accepting landlessness as a posture of faith. . . . The sojourn is freely chosen, not imposed. *It is a choice made by those who could have chosen not to leave.* The choice means to throw one's self totally on Yahweh, not in order to live in some nonhistorical relation with God but to be led to a better place, one characterized by promise not known either in Ur or in one's father's house.[14]

I am learning that I, if I want to be fully Holly, must accept my spot as a sojourner right alongside Aisha.

"Sojourner" is such a heroic word. It's not much of a stretch for us to get up out of our beach chairs and sojourn, because it sounds gallant, brave even. If we say we're sojourners, it is almost with the understanding that we will be admired for what we do. And while biblical translators have used that word to express these ideas of locality, the truth is that sojourners of faith are more like resident aliens. They're refugees, like my friend Aisha. Most people who encounter her don't see her as brave. They wonder why she covers her head, why she wears sandals in the winter, and why she can't do anything but squint at them whey they speak English to her. She sometimes fumbles with money and sometimes holds her books upside down. She looks and acts twenty years older than she really is because of her lifelong lack of access to medical care. She is the epitome of a resident alien. She is what it means to be a sojourner.

If we're honest with ourselves, we don't like that definition. We want a place. We don't want to be homeless, because homeless people are suspect. We don't want to be resident aliens or go without places to lay our heads. We don't want to look silly or incompetent. A friend who lives in Turkey told me what this resident-alien struggle was like for him during the first few months of living there and learning the language. It was hard, he said, to be willing to sound like a preschooler each time he tried to communicate with someone. He said it made him want to make a t-shirt he could wear that said, "In my own culture, I have a graduate degree." Another friend said the worst part of transitioning to a new culture was hearing people speak to their pets and realizing that the pets understood more of the language than he did. We don't like this pride-swallowing, sacrificial sojourning as much, do we? We like to look competent and confident, like God has helped us "get it all together." He has, of course, but the Christian sojourn is not a Carnival cruise to a new life. It is a daily walk of faith that is reaching out in hope of things unseen. It is one that is willing to identify, figuratively and literally, with the resident aliens of our world.

Brueggemann writes poignantly on this idea as it relates to our churches here in America: "It is likely that our theological problem in the church is that our gospel is a story believed, shaped, and transmitted by the dispossessed; and we are now a church of possessions for whom the rhetoric of the dispossessed is offensive and their promise is irrelevant. And we are left to see if it is possible for us again to embrace solidarity with the dispossessed."[15]

I think we should try it. I think we should fold up our beach chairs, reject the fruitless pursuit of space, and live deeply in our collective places. Let's begin to walk in faith, not worrying about looking silly, not worrying that we don't see the end of the road God has asked us to travel. I think we should do this—not because it's our duty or because the Bible says we should do it—but because the walk itself is where we'll truly be at home. We should try it because we read in Scripture about the great reversal, the first being last and the last being first. We walk together, knowing this place is not all there is. We walk with the conviction that "grasping for home leads to homelessness and risking homelessness yields the gift of home."[16]

Discussion Questions

1. How much time in each day do you spend engaged in your place? How much time do you spend in "non-place"?

2. How do you "embrace solidarity with the dispossessed"? How does your faith community do this?

Notes

1. John Inge, *A Christian Theology of Place* (Hampshire, Great Britain: Ashgate Publishing Ltd., 2003) 92.

2. Names and details have been changed here and in other chapters out of respect.

3. Marc Augé, quoted in Philip Sheldrake, *Spaces for the Sacred* (Baltimore MD: Johns Hopkins University Press, 2001) 8–9.

4. Ideas from Augé here are from a helpful overview found in Sheldrake, *Spaces for the Sacred*, 8-9.

5. Walter Brueggemann, T*he Land: Place as Gift, Promise, and Challenge in Biblical Faith*, 2nd ed. (Minneapolis: Fortress Press, 2002) 200.

6. Ibid., 4.

7. Ibid.

8. Ibid.

9. Ibid., 200.

10. Colossians 1:27.

11. Quote from Brueggemann in Inge, *A Christian Theology of Place*, 45.

12. This idea is discussed at length in Inge, *A Christian Theology of Place*.

13. Brueggemann, *The Land*, 20.

14. Ibid., 6.

15. Ibid., 206.

16. Ibid., 202.

Chapter 3

Coming Near

If I see my city as beautiful and bewitching, then my life must be so too. —Orhan Pamuk[1]

When we examine the idea of place, we need to explore the ideas of *near* and *far*. We first begin to understand the two concepts as children. The words *near* and *far* are a way to describe our place in relationship to that of another. Think *Sesame Street*. Remember Grover, running up to the front of the television screen and then running so far back he was almost out of view? "Near . . . far . . . near . . . far." And so our first understanding of these two ideas was a way to categorize physical space.

As we continue to live, grow, and gain new experiences, we sense that these two terms are broader and describe more than physical space. A friend moves away, but we notice how things always "seem the same" when we talk on the phone together. We might have a difficult relationship with the person on the other side of the apartment wall. We notice that our high number of Facebook friends doesn't necessarily fill us up the way a cup of coffee and a good conversation do. Life experiences continue to teach us more about these ideas of locality: we begin to see that our understanding of nearness and our ideas of distance might need to stretch. Distance, we learn, is not only about miles. And nearness is not just about living in the same place.

We feel this in our spirits because, though we live at a time when technology has given us the ability to shrink miles, we still don't automatically feel "settled." We can be everywhere but feel as though we're nowhere. "The frantic abolition of all distances brings no nearness," says philosopher Martin Heidegger, "for nearness does not consist in the shortness of distance."[2]

Even the pig named Wilber from E. B. White's *Charlotte's Web* understands this dilemma of place. Watching the movie version recently with my kids, I enjoyed the scene where Wilbur is getting to know his fellow barn animals. "You're all friends, right?" he asks a goose. "Oh, sure," the goose answers, "we've been here together our whole lives." What Wilbur says always pricks me: "I'm not so sure being in the same place is the same as being friends."[3]

Whether we are far from a loved one and want to be near or if we are too close to someone for comfort, it all points to our desire for authentic connection with a place and the people in it. We want to matter; we want to be missed when we're gone. Heidegger says, "What is it that unsettles and thus terrifies? Namely . . . that despite all conquest of distances the nearness of things remains absent."[4]

My husband and I have had the chance to travel quite a bit over the past ten years, and, in some ways, because of his work, we're probably now seen as the people with airline tickets in their pockets, ready to put others on a plane at a moment's notice. Some people like that about us, and some people run and hide from us. And probably more than any theological or biblical quibble, we find ourselves talking with other people about what I'd call a philosophy, or even a theology, of travel. We are talking about *near* and *far*.

As we recognized before, *near* and *far* have countless meanings. For some, travel is glamorous, evoking memories of a time when people wore suits or dresses and heels to the dining car of a train. For others, travel is a necessary part of work: many get up before dawn, are in Dallas by 8 a.m., and arrive back home by dusk to tuck the kids in. *Near* and *far* sometimes mean a vacation. Sometimes it means deployment. And let's not forget that for many people in the world, the possibility of travel past where they could walk is a privilege never once granted them.

What's more, there are many motives for travel. Some people save religiously for the once-in-a-lifetime pilgrimage. Some collect airline miles or stamps on their passports, feeling that they are achieving something by seeing new things. Some go to learn, others to give and do. Others feel that waking up in a new setting will allow them anonymity or the ability to be people they are incapable of being in everyday life. Some, who have talents and skills that are needed in far-off places, go to help other people. And aren't there times when we just need to *go*? Don't we need to escape? To avoid the routine of daily life?

Further still, there are many reasons *not* to travel. It is *work* to travel; preparing to be away is often more work than staying home. Travel is hard on your body and hard on your spirit. Whether traveling for pleasure or for work, you miss people and the particulars of your home. It is a sacrifice to be away. When you travel, there is discomfort, inconvenience, and uncertainty.

There is legitimate fear in the threats of travel, from bedbugs to bombers. Many times there is no space, in our schedules or in our minds, for travel; the full agenda we have at home, from soccer games to sicknesses, seems like all we can or want to handle.

I write this to say that travel is not a magical experience. Whether you travel for selfish or unselfish reasons, there is no sprinkling of fairy dust, as they say, when you cross an ocean. What I've found is that miles are irrelevant; if we let life roll by us at home, Venice will probably roll by us in the same manner.

Near is a choice. *Far* is a decision. Wherever we are, there is a necessary intention we must have that allows us to experience meaningful connection to people and places. We know this to be true in our hearts from life experience, don't we? There are people who travel quite a bit but seem to be hermits in their spirits. There are those who have never traveled outside their county—but when you are with them, you sense the wide-open spaces in their hearts. We find that *near* and *far*, while helpful in categorizing relative distances, don't necessarily capture the essence of connection with a place and its people. What makes something *far*, in this sense, is avoiding it. What makes someone *near* is our choice to embrace him or her.

God created us for authentic connection to places and people. He asks us to come near. We must make the choice to be present and to be a loving spirit regardless of where on the planet we are, where we thought we'd be, or even where we *want* to be. This adventure isn't anything God is unwilling to do himself. He is Immanuel, *God with us*. He is the God who "became flesh and blood and moved into the neighborhood," as it says in John 1:14. Christ moved among us with intention. He was willing to walk through Samaria and he was willing to go back to his hometown. In order to follow his lead, in order to be the kind of humans he created us to be, we must embrace our places with intention. It doesn't matter where we are; what matters is that we come close to our place and its people.

This takes intentionality. Consider Wendell Berry's thoughts:

> One of the peculiarities of the white race's presence in America is how little intention has been applied to it. As a people, wherever we have been, we have never really intended to be. The continent is said to have been discovered by an Italian who was on his way to India. The earliest explorers were looking for gold, which was, after an early streak of luck in Mexico, always somewhere farther on. Conquests and foundings were incidental to this search—which did not, and could not, end until the continent was finally

laid open in an orgy of goldseeking in the middle of the last century. Once the unknown of geography was mapped, the industrial marketplace became the new frontier, and we continued, with largely the same motives and with increasing haste and anxiety, to displace ourselves—no longer with unity of direction, like a migrant flock, but like the refugees from a broken anthill. In our own time we have invaded foreign lands and the moon with the high-toned patriotism of the conquistadors, and with the same mixture of fantasy and avarice.[5]

No wonder we feel displaced; we come by it naturally. We're not intentional about our places, it seems, because we're always on the way to "somewhere else." Consider how often we slight the present. We breeze by the clerk at the checkout counter because we have two other stops to make before home. We minimize a child's sadness about a relationship at school because it's almost bedtime. It can happen easily, and life is tricky here: the inability to be present can mask itself as ambition, which is such an admired trait. We're successfully crossing off items on our "to-do" lists, and before we realize it, we're not *near*, we're *far*.

If we live like we are constantly on our way somewhere else, where *are* we? Recently, Ruben Bruno Hernandez, a friend to our family, told me a story about this idea. Ruben is an accountant for a large produce company in the south of Spain. He also works bi-vocationally to direct a network of churches that aid African immigrants living there and working in the greenhouses. Ruben was driving home from work one Friday night after a long week. He was tired, it was after dark, and he was supposed to meet a group of young African friends who live a ways from his house. He'd postpone it, he thought. "I'm tired. I'm just ready to go home." He skipped the exit to his house, however, and made the forty-five-minute drive toward the home of his friends. When he arrived, he was welcomed warmly. The men had been waiting there for some time for his visit, and he was greeted with a large meal that had been prepared at great expense. His visit went on to refresh both his African friends and himself. What if he hadn't gone? It is so easy to fill our calendars with activities that tire us, which limits our ability to be present and deepen our relationships with others. We must not let time or energy keep us from drawing near to a place or its people. We must listen and, as Ruben has taught me, be obedient to the voice of God in each circumstance.

Neither must we let diminishing time be a factor. It is easy for us to "check out" in a place or in a relationship if we know change is coming. For example, in our last two months of living in South Africa, a friend of ours recommended we join a small Bible study group at our church that was active in serving the HIV community. Since we were also serving in this area, he thought it would benefit us to connect with people who were thinking through similar issues. Matt and I then had a choice: should we join a new group with just two months left? We could come up with a lot of reasons not to bother. We should savor the time . . . we're going to be traveling one of those weeks anyway . . . we don't really care for the particular book they're studying . . . we don't even know these people . . . why get involved when we're going to be leaving soon? We'd made a decision to squeeze every last drop from our diminishing time there, however, so we ended up trying it. I can still remember some of the formative conversations that took place in the homes of those hospitable friends. What if we hadn't gone? What if we'd missed those friends? Those words? We have to learn to live in the present and remember to live with intention, even when there may be just moments left.

Sometimes brilliant things don't come with our intention. Sometimes this day-in, day-out intention is simply work. It is going to meetings you don't want to go to. It is getting out and visiting a sick friend when you'd rather snuggle up at home. We all seem to want the closeness and the meaning in our lives that comes with intention, but are we willing to do the work of interaction? Sometimes it is in the work of "showing up" that the shining moments come. We must work toward becoming disciplined, reliable people. Why? Because closeness to a place, to people, doesn't come automatically, even if you live in a place or with a person your whole life. Instead of withdrawing or deciding not to follow through, we must work to be in authentic relationships with others. When we find ourselves in the give and take in a place and with a people, we will find ourselves *near* instead of *far*.

We must do the work of knowing our place. This means we need to know our town's idiosyncrasies, the makeup of its people, its traditions. We need to be proud of its randomness and appreciate its struggles. For example, you probably didn't know this, but my hometown boasts the fine distinction of hosting the world's shortest St. Patrick's Day parade: at 9 a.m., it begins on one side of the street at a soda fountain and ends on the other side of the street at a local bar. Or maybe I should let this quote by Scott Russell Sanders say it much better:

> It is rare for any of us, by deliberate choice, to sit still and weave ourselves into a place, so that we know the wildflowers and rocks and politicians, so that we recognize faces wherever we turn, so that we feel a bond with

everything in sight. The challenge, these days, is to be somewhere as opposed to nowhere, actually to belong to some particular place, invest oneself in it, draw strength and courage from it, to dwell not simply in a career or a bank account but in a community. . . . Once you commit yourself to a place, you begin to share responsibility for what happens there.[6]

Geographer and philosopher Yi-Fu Tuan calls this idea *topophilia*, which, put simply, means place-love. When we commit ourselves to this kind of consciousness about where we are on the planet, it is an active pursuit. "Ask questions," it says in Deuteronomy 4:32. "Find out what has been going on all these years before you were born." We make sure that we find beauty in the familiar. We engage the people and places around us, putting effort and energy into attention.

When we work at place-love, we discover a sense of informed contentment. We aren't living in a blind nostalgia of idealism. Instead, we work at observing and understanding others. We withhold snap judgments and seek to understand alternate perspectives. In this sense, we find that we are changed for the better as well. In talking about one's love of a place, I have to include the following passage from another of Wendell Berry's essays. Once again, he captures this idea, in a discussion of a farmer and his farm, much better than I can:

> It has not been uncharacteristic for a farmer's connection to a farm to begin in love. This has not always been so ignorant a love as it sometimes is now; but always, no matter what one's agricultural experience may have been, one's connection to a newly bought farm will begin in love that is more or less ignorant. One loves the place because present appearances recommend it, and because they suggest possibilities irresistibly imaginable. One's head, like a lover's, grows full of visions. One walks over the premises, saying, "If this were mine, I'd make a permanent pasture here; here is where I'd plant an orchard; here is where I'd dig a pond." These visions are the usual stuff of unfulfilled love and induce wakefulness at night.
>
> When one buys the farm and moves there to live, something different begins. Thoughts begin to be translated into acts. Truth begins to intrude with its matter-of-fact. One's work may be defined in part by one's visions, but it is defined in part too by problems, which the work leads to and reveals. And daily life, work, and problems gradually alter the visions. It invariably turns out, I think, that one's first vision of one's place was to some extent an imposition on it. But if one's sight is clear and if one stays on and works well, one's love gradually responds to the place as it really is,

and one's visions gradually image possibilities that are really in it. Vision, possibility, work, and life—*all* have changed by *mutual correction.* Correct discipline, given enough time, gradually removes one's self from one's line of sight. One works to better purpose then and makes fewer mistakes, because at last one sees where one is. Two human possibilities of the highest order thus come within reach: what one wants can become the same as what one has, and one's knowledge can cause respect for what one knows.[7]

When we take the time to understand our locality, to work at fully dwelling in our place, there is mutual correction. As Berry says, our worldview is stretched and we all change for the better. We must fully know our place, acknowledging all quirks, faults, and bright spots, for that process, in turn, helps us understand ourselves and our world. "One place comprehended," says Southern fiction writer Eudora Welty, "can make us understand other places better."[8]

When we lovingly and realistically embrace our place, Christ helps us to be fully present. We come close to people and places. In doing so, we remind ourselves and others that God is a loving, immanent God, a God who came—and continues to come—*near.*

Discussion Questions

1. To what or to whom do you feel *near* in your life?

2. How have you woven yourself into your place and shared responsibility for what happens there? How might God be leading you to come near to the people and places in your life?

Notes

1. Orhan Pamuk, *Istanbul* (New York: Alfred A. Knopf, 2006) 56.

2. Martin Heidegger, quoted in John Inge, *A Christian Theology of Place* (Hampshire, Great Britain: Ashgate Publishing Ltd., 2003) 13.

3. "Wilbur Tries to Make Friends," *Charlotte's Web*, DVD, directed by Gary Winick (Hollywood: Paramount, 2006).

4. Heidegger, quoted in Inge, *A Christian Theology of Place*, 13.

5. Wendell Berry, *The Art of the Commonplace: The Agrarian Essays of Wendell Berry* (Washington, DC: Shoemaker & Hoard, 2002) 35.

6. From the essay "Local Matters" by Scott Russell Sanders, quoted in bell hooks (pen name intentionally lowercased), *Belonging: A Culture of Place* (New York: Routledge, 2009) 67–68.

7. Berry, *The Art of the Commonplace*, 186–87, italics mine.

8. Quoted from Welty's 1956 essay, "Place in Fiction," used in Melissa Holbrook Pierson, *The Place You Love Is Gone: Progress Hits Home* (New York: W. W. Norton & Company, 2006) 190.

Chapter 4

Coves and Kayaks

The first river you paddle runs through the rest of your life. It bubbles up in pools and eddies to remind you who you are.[1]

There is a place where my family likes to go to get away. It is a quiet, Baptist place (I know: it sounds like an oxymoron). There are cabins on a lake, and I've made visits there all my life. Sometimes I went several times a year, and at other times a decade passed between trips. It is simply a place to rest and be with God. Summer and winter, we've gone with friends and family, and we've also gone alone. It's a place that's not far from where we live and at the same time far enough.

There is a sense in which my visits are nostalgic. I think about going to music camp there or when a woman from our church would take a van load of giggly girls down for an overnight getaway. When I see the hilly line of the trees against the sky in the cove, my mind replays the hours of childhood videos we have of us and our friends skiing and tubing in the same cove with the same backdrop. I get a sense of satisfaction as I walk hand in hand with my children on the docks or swing with them in the creaky wooden swings by the lake. It's fun to stand with my kids at the mouth of the cave, to peer in and see myself as a child, also afraid to enter.

On a recent trip there, however, I was reminded that there is so much more to a place than nostalgia. My husband and I got up early one morning to take our kayaks out. As I paddled around the corner of the cove, I had a total girl thought: *I wonder how many hairstyles I have had since I started coming here.* I've had fine baby hair, braids, sticky 1980s bangs, pixie cuts, and now gray hair. And here this cove is—the trees, the water—here the whole time, looking the same year after year. I've sucked a pacifier here, tight-rolled my jeans here, and changed my own kids' diapers here, and all the while this cove remained the same. I had elementary school programs, endured high school dating dramas, and chose a college while this cove was here. I started my first job, got married, and moved to Africa, and all the while the cove was still here. I finished school and added two children to my life. The cove was still here. It was almost as if the cove existed outside of time, while the minute details of my life washed up on its shore that morning.

God reminded me there in the kayak that, even as long as that cove had been there, it really wasn't about the cove. As timeless as it seemed that morning, the cove, too, will change. It wasn't about locality as much as it was about something even more timeless: connection with God himself. The cove wasn't the place. God was the place.

Many times, faith tries to make place definitive. Think about it: Jerusalem. Mecca. A compound in Waco. A temple in Lhasa. A river in India. The danger of loving the particulars of our places is our tendency to become exclusive, to put a wall up around our place and label it "ours."

Jesus speaks to this tendency in John 4. He meets a woman at a well in Samaria. Simply the fact that he's chosen to walk through Samaria instead of walking around it like all the "good Jews" gives us a glimpse into God's perspective on locality; Jesus does not allow place and the associated religious and political debates to get in the way of his connection to a person. As the woman finds out who he is, she begins her religious chitchat about the right place to worship. Can you imagine what Jesus is thinking? Here is God Incarnate sitting on the well before her, engaging her and the details of her life, and she wants his perspective on the latest denominational controversy: should we worship here on our mountain or there in Jerusalem? And just as there is so much more than nostalgia about my cove, Jesus gently helps the woman to understand that there is more to worshiping God than latitude and longitude. He tells her that he himself is the living water. He himself is the place to which she must go.

We can't be too hard on this Samaritan woman. It's as if Jesus tells the woman at the well, "Well, you're missing the point. It *is* about place, but it's even *more* about me, in this place, *here with you.*" And if we read carefully, it seems that the biblical account retells this same story—that locality is important but not definitive. Over and over, we read stories of people like us who need this shift in focus.

We begin with the garden stories, the ones about how the perfect dwelling place wasn't one where every wish was fulfilled but where relationship with God was the ultimate fulfillment. We read of Abraham choosing that relationship with God over his own place and his own family. We read in Exodus of a whole people leaving Egypt behind so that they could worship God. We see God asking them to build the tabernacle in Exodus 25:8 (NIV): "Then have them make a sanctuary for me, and I will dwell among them." We read chapter after chapter of the ways in which this dwelling place for God was revered, but was it really about the place itself, about the dyed cloth, the gold instruments, and the ark of the covenant? Leviticus 26:11-13 gives us more of an explanation: "I'll set up my residence in your neighborhood; I won't avoid or shun you; I'll stroll through your streets. I'll be your God; you'll be my people. I am GOD, your personal God who rescued you from Egypt so that you would no longer be slaves to the Egyptians. I ripped off the harness of your slavery so that you can move about freely." Instead of place being definitive here, we see a God who is trying to reset our priorities, one who is sitting before Israel as he later sat in front of the Samaritan woman. "It *is* about place, but it's *more* about me, in this place, here with you."

Walter Brueggemann says it like this: "The land for which Israel yearns and which it remembers is never unclaimed space but is always *a place with Yahweh*, a place well filled with memories of life with him and promise from him and vows to him."[2] It makes sense, doesn't it? For if it was really about the promised land, would they have received manna in the wilderness? If it was really *only* about place, wouldn't all the provisions have come once they reached the destination? We see the opposite, don't we, when we read about the ways God led the people in a cloud and in a pillar of fire? We see him providing water and manna, trying to help the people understand that the covenant with him was more important than the goal of a destination. "Israel is shown that life-giving resources do not come from land but from Yahweh," says Brueggemann. "Israel is not tied to, dependent upon, or subservient to the land. The Lord of chaos gives these resources to the landless."[3]

The story of this people progresses, and the convulsive people in the wilderness grow into a united kingdom, settled, powerful, and wealthy. One day, it occurs to Israel's king, David, that his own place is finer than the dwelling place of God. "Here I am, living in a palace of cedar, while the ark of God remains in a tent," reads 2 Samuel 7:2 (NIV). God, through the prophet Nathan, again tries to guide David's perspective:

> I have not dwelt in a house from the day I brought the Israelites up out of Egypt to this day. *I have been moving from place to place with a tent as my dwelling.* . . . Now then, tell my servant David, "This is what the LORD Almighty says: I took you from the pasture and from following the flock to be ruler over my people Israel. *I have been with you wherever you have gone,*

and I have cut off all your enemies from before you. Now I will make your name great, like the names of the greatest men of the earth. And *I will provide a place for my people Israel.*"[4]

Again, we see the recurring story of a God who is willing to come find his people, even if it means plucking us from a sheep-covered hillside, in order to be in relationship with us. The point, he reminds David, is not that God dwells in a specific place but that *God* wants to be in *our* specific places with *us*. David is astonished. "Is this your usual way of dealing with man, O Sovereign LORD?" he asks (2 Sam 7:19, NIV).

And we see, as we continue to read the words of David, that it is. "Lord, you have been our dwelling place throughout all generations," he writes in Psalm 90:1-2 (NIV). "Before the mountains were born or you brought forth the earth and the world, from everlasting to everlasting you are God." *The Message* puts verse 1 like this: "God, it seems you've been our home forever."[5] In the next psalm, David reminds us that because God is with us, nothing is worthy of our fear:

> Fear nothing—not wild wolves in the night, not flying arrows in the day, not disease that prowls through the darkness, not disaster that erupts at high noon. Even though others succumb all around, drop like flies right and left, no harm will even graze you. You'll stand untouched, watch it all from a distance, watch the wicked turn into corpses. Yes, because GOD'S your refuge, *the High God your very own home,* evil can't get close to you, harm can't get through the door.[6]

We see that David, the man after God's own heart, finds his ultimate comfort in the fact that God's presence is with him always.

> I look behind me and you're there, then up ahead and you're there, too—your reassuring presence, coming and going. This is too much, too wonderful—I can't take it all in! Is there anyplace I can go to avoid your Spirit? to be out of your sight? If I climb to the sky, you're there! If I go underground, you're there! If I flew on morning's wings to the far western horizon, You'd find me in a minute—you're already there waiting![7]

So David does not build God a place but relishes the fact that God's place is within us. He makes plans for a temple and tries to hand down the legacy of this knowledge to his son, Solomon, who eventually becomes king. And Solomon does build. "I have indeed built a magnificent temple for you, a place for you to dwell forever," we read in 1 Kings 8:12-13. In the same breath, Solomon questions,

> Can it be that God will actually move into our neighborhood? Why, the cosmos itself isn't large enough to give you breathing room, let alone this Temple I've built. Even so, I'm bold to ask: Pay attention to these my prayers, both intercessory and personal, O GOD, my God. Listen to my prayers, energetic and devout, that I'm setting before you right now. Keep your eyes open to this Temple night and day, this place of which you said, "My Name will be honored there."[8]

Could the king have been thinking of Eden? For even as he dedicates the temple, one that probably exhausted all the possibilities of earthly architecture, Solomon doesn't glory in the temple itself. Instead, he pleads for God's presence to be with him and his people.

It's the same pleading we hear in the voice of the prophet Isaiah. He is pleading, however, with the people to recognize the presence of God and to let go of their preoccupation with their place and their power. The message he gives to them in Isaiah 66:1-2 is what God has said to him: "Heaven's my throne, earth is my footstool. What sort of house could you build for me? What holiday spot reserve for me? I made all this! I own all this . . . but there *is* something I'm looking for: a person simple and plain, reverently responsive to what I say." God is looking for a place in us, Isaiah poignantly tells them. If he finds it, says Isaiah, if we reshuffle our priorities and find that God is our place, we will have found our Eden.

> Just take a look at Zion, will you? Centering our worship in festival feasts! Feast your eyes on Jerusalem, a quiet and permanent place to live. No more pulling up stakes and moving on, no more patched-together lean-tos. Instead, GOD! GOD majestic, *God himself the place* in a country of broad rivers and streams, but rivers blocked to invading ships, off-limits to predatory pirates. For GOD makes all the decisions here. GOD is our king. GOD runs this place and he'll keep us safe.[9]

Ezekiel, another prophet of God, sees this same idea at the heart of God. He witnesses God's desire for a place with us as a valley of dry bones comes to life and as two sticks of Judah and Israel are held together and brought back from exile. He hears God remind him of David, and how he shepherded the people not just into a temple but into a knowledge of God. He hears God foretell his honoring of his covenant relationship with humanity once again:

> My servant David will be their prince forever. I'll make a covenant of peace with them that will hold everything together, an everlasting covenant. I'll make them secure *and place my holy place of worship at the center of their lives forever.* I'll live right there with them. I'll be their God! They'll be my people! The nations will realize that I, GOD, make Israel holy when *my holy place of worship is established at the center of their lives forever.*[10]

And through all the measuring of the temple, through all the inspection, we read the final word from Ezekiel: "From now on the name of the city will be YAHWEH-SHAMMAH: GOD IS THERE."[11]

Is it any wonder, then, with this rich history of place and purpose, that God explained the birth of his son with the words *Immanuel, God is with us?* And as we look at accounts of the life of Christ—all the stories, all the people, all the places—what do we see? Do we see a priority of place? Do we see him born in a palace? Does he hang with the powerful? Do we see him vie for the head seat at the table? Do we find Jesus prizing place? No. Above all, we see him prizing people.

We find him well side in the heat of the day, with the Samaritan woman. We find him announcing that God wants to have his place in the hearts of humanity. And, eventually, we see him crucified for it. We read of Stephen in the book of Acts, and how he was stoned for his sermon on the proper perspective of tabernacle and temple. Although the people around him couldn't see it, Stephen could see way off in the distance what the book of Revelation foretells: "There was no sign of a Temple, for the Lord God-the Sovereign Strong-and the Lamb are the Temple." The real meaning of all this, says the author of Revelation, is this: "Look! Look! God has moved into the neighborhood, making his home with men and women! They're his people, he's their God. He'll wipe every tear from their eyes. Death is gone for good—tears gone, crying gone, pain gone—all the first order of things gone."[12]

Is place important? Yes. Is place definitive? No. The biblical accounts teach us that life is not so much about place as it is that *God is with us in our own places.* He is willing to be with us, in our own context, at this particular time in history. And that—as we read in the lives of Abraham, David, Ezekiel, and Jesus—changes everything. It changes my approach to my place. It changes me to know that through all the hairstyles, through all the good and yucky stages in my life, God has been with me, has *wanted* to be with me. The grace of that knowledge challenges me to remember that my true place is ultimately with him. And apparently, I do need to get away, to get out in a kayak and paddle around a cove, to remember that.

Discussion Questions

1. Are there specific places in your home or specific places in your life where you go to be with God?

2. What difference does it make to you to know that God is with you in your place today?

Notes

1. Lynn Noel, http://www.oklahomaroadtrips.com/Paddler-Quotes.htm, (accessed 29 July 7 2011).

2. Walter Brueggemann, *The Land: Place as Gift, Promise, and Challenge in Biblical Faith*, 2nd ed. (Minneapolis: Fortress Press, 2002) 5.

3. Ibid., 30.

4. 2 Samuel 7:6, 8-10, italics mine.

5. Psalm 90:1.

6. Psalm 91:5-10, italics mine.

7. Psalm 139:5-10.

8. 1 Kings 8:27-29.

9. Isaiah 33:20-22, italics mine.

10. Ezekiel 37:25-28, italics mine.

11. Ezekiel 48:35.

12. Revelation 21:3.

Chapter 5

The Late One

He is the true American pioneer, perfectly at rest in his assumption that he is the first and the last whose inheritance and fate this place will ever be.[1]

Have you ever been in a meeting and watched someone walk in late? Sometimes she might sneak in the back with her cup of coffee and calendar, trying to go unnoticed. Other times she tries to overcome it with a grandiose entrance, rushing in with a flurry of file folders. She tells long, loud tales of woe, and sometimes food is involved in the interchange. However it happens, we usually listen politely for a moment and wait for the meeting to settle back into the topic at hand. Somehow it seems rude to ask how a person who ran into so many different evils on her commute had time to stop for breakfast, so you sit and munch in silence. After all, you are all gathered for a common purpose, and there is an agenda for this meeting.

In this particular situation, have you ever gone on to observe the person who walked in late—let's call her "the Late One"—begin to hijack the meeting? It's one thing for her to walk in so late, but she brought breakfast burritos and we forgave her. Now, though, she begins to interject her ideas loudly, even though she hasn't heard the previous thirty minutes of discussion. You are annoyed because now time is going to be wasted apprising the Late One of what the others at the meeting have already decided. At the same time, you pity the Late One. She is embarrassing herself because she's not recognizing anything took place prior to her arrival.

Recently, I was reminded that many times I am the Late One. Maybe I literally walk into a situation with our church and act as though my entrance, or my generation's entrance, is where it all began. Or I might be in a conversation with a friend and quickly jump to, "Well, I think this is what you should do" before I really listen to the history of the situation. Sometimes I am guilty of seeing a problem in our city and assuming my answer is *the* answer. Many times, I am the Late One because I rush into prayer spouting life details that "prevented" me from being where God already asked me to be.

What I'm learning is that we need to guard the way we approach this life of faith. Cognitively, we realize that it is not all about us, but sometimes our actions prove that we view life from a fairly self-centered perspective. If we as Christ-followers are not careful, we begin trying to run the meeting instead of being humbly mindful of our lateness. It is almost as if we come humbly to God for salvation and claim his action for it, and then immediately set to work charting our own course for a holy life for ourselves and others.

We forget that the God who created everything and set the world in motion is still creating new life in nature and in people. We forget that we are just a few of the many people throughout time and across the world that have called on God in need of his grace. We forget that the one who planned the universe and planned to draw us to himself is still planning, still reconciling people to himself. He is in control. In years past, we've sung about it like this: "This is my Father's world, O let me ne'er forget / That though the wrong seems oft so strong, God is the Ruler yet."[2]

We need to be aware of our place. It is a place behind God, one where we follow Christ because he goes before us. It is a restful place. It is a joy-filled, simple place. He is charting the course; it is our job to let the Spirit lead us on it. We are joined together by God's loving initiative and for his purposes. He is the one who is graciously running this meeting; I certainly don't want to be the Late One. I need to remember that much has taken place prior to my arrival and that I would be wise to listen first. Thankfully, the invitation to the gathering includes me, even when I am late. I only make it awkward and frustrating for others (and for God, I imagine) when I act as though everything revolves around me.

I am learning how right and how freeing it is to lose my self-referential ideas about faith and look instead to the ongoing, graceful leadership of God throughout time. Eugene Peterson writes about this "Late-to-the-Meeting" idea in *The Contemplative Pastor*, saying, "We misunderstand and distort reality when we take ourselves as the starting point and our present situation as the basic datum."[3] There is so much more than what we see at present, and how exciting that knowledge is!

The whole of Scripture confirms God's hospitable initiative in inviting us to the table. It reminds us that though our place at the table is valued, it is not central. From its inception, we read "In the beginning, God . . . ," and from that point on, we read of a God who is constantly spoken of as "one who goes before." We read of a God who has all authority, who orchestrates life in a way that is true and right. The reality is that God is and has been and will be working. There is an agenda, and there are things to accomplish. "We will always be living a mystery—but a good mystery," writes Peterson in *Christ Plays in Ten Thousand Places*.[4] This knowledge doesn't require us to abandon the mystery but simply to realize that *we* aren't perpetuating it. And that's a good thing.

God goes before. We read it in Genesis when he decides to pluck Abraham and his family from obscurity and make a family for him. We read it in Exodus when God manifests himself in a pillar of cloud and fire in order to lead the Hebrews. We read it in Deuteronomy 9:3 when Moses says, "Today know this: GOD, your God, is crossing the river ahead of you." We read it in Isaiah when God promises to go before the people in order to make the rough places plain (Isa 40:4, KJV). We hear Jesus say, "After I have risen, I will go ahead of you," and "I am going to prepare a place for you" (Matt 26:32, NIV; John 14:2, NIV). We read in Paul's letter to the Romans that God went ahead in loving action and gave his son for us, even while we were sinners. In Romans 5:6, he writes, "Christ arrives right on time to make this happen. He didn't, and doesn't, wait for us to get ready. He presented himself for this sacrificial death when we were far too weak and rebellious to do anything to get ourselves ready." Scripture is filled with stories not about man's efforts to get to God but about God's anticipating action and preemptive pleasure.

We read in Colossians 1:17 (NIV) that Christ is "before all things, and in him all things hold together." *The Message* version of this part of Colossians reads like this:

> He was supreme in the beginning and—leading the resurrection parade—he is supreme in the end. From beginning to end he's there, towering far above everything, everyone. So spacious is he, so roomy, that everything of God finds its proper place in him without crowding. Not only that, but all the broken and dislocated pieces of the universe—people and things, animals and atoms—get properly fixed and fit together in vibrant harmonies, all because of his death, his blood that poured down from the Cross.[5]

According to what Paul writes here, when we choose to follow Christ, we don't just take a spiritual bath, get cleansed of sin, and ensure a pleasant afterlife. When we begin to follow Christ, we are deciding to sit down at the table, where the task at hand is life and more life. It is the decision to let

God's goodness and rightness begin to open up your life, and to help that wholeness reach others as well. It is the decision to join a work already in progress, the work of God's peacemaking.

God has been working and recreating to bring wholeness all through history. Think about it this way: flood, Davidic monarchy, exile, and God was there. Constantine, Chinese dynasties, plague, and God was there. Colonialism, slavery, industrialization, wars, and God was there. Holocaust, civil rights, and apartheid, and God was there. Modern-day hurricanes, tsunamis, earthquakes, droughts, and God is there. Child soldiers, human trafficking, torturous dictators, and God is still working. He has been present and working to right human wrongs for century upon century. Who are we? The Late Ones. Instead of rushing into the meeting with our religious plans on how to help ourselves or help the world, we need to come humbly to the conference table with an accurate understanding of our place. Our feeble attempts at strategy will not eradicate the evils of the world. Instead, the humble acknowledgment of our place—our exercise of trust—will help further God's loving work in the world.

Sometimes, the practice of place is in the small, simple things. I was frustrated about a recent situation and took a few moments to pray about it. The word I heard back from the Lord was clear. It was a simple yet kind: "I've got this." And by going back and recognizing his action, I realized I needed to move forward with this knowledge. The situation did not change after my prayer; my perspective did. I could rest in the knowledge that God is the one going before and that he's "got this." That reminder has changed the way I bring other requests to God as well. Instead of saying "God, could you . . . ?" or "God, would you . . . ?" I begin with, "Lord, you've got this . . . and you've got that . . . ." He does. I've always believed he does, but using these words has been a simple way to acknowledge that he is going before me and orchestrating every situation I might mention.

My responsibility goes further than sitting at the conference table, however. It is not only my recognition of God's prevenience, his "going before," that helps to make things right in the world; it is my willingness to follow. I was asked to this meeting for a reason, and that reason involves but is not limited to my own fulfillment. Once again, it's bigger than me. I was created and trained to be me with a specific purpose in mind. There are ways God could use me in helping to make the world right, and I will leave this meeting with a list of tasks. These are not chores—I am especially equipped to do them—but they *will* ask a lot of me. If I recognize my place and remember that God is going before me, I become who I am meant to be. More important, something that was wrong in the world has been made right.

Last winter, my family and I decided to go sledding. I went to the basement to gather the coats, hats, mittens, and boots for the four of us. I put these by the garage door before I went back upstairs, snow pants in hand, to help begin the layering. Matt and Lucy were dressed and ready for fun. They scampered down the stairs as my son Mikias (pronounced mih-KEE-us) and I began to follow at a slower pace. We were clumsy in our layers. At the top of the stairs, Mikias stopped with a troubled look on his face.

"Mommy," he said, frowning, "I don't have my mittens."

"Let's walk down the stairs," I told him, holding his hand. "Everything you need is down there."

"But *Mommy*," he said again. His face said, "Clearly, she doesn't get it," and his mouth said again, "I don't have my *mittens*."

"Yes, I know, Mikias. Just walk down the stairs." By this time my husband, daughter, and our sleds were in the car already.

"*Mommy!*" he said once again. "I. Don't. Have. My. *Mittens!*"

And during our descent (probably because of the excruciating length of time it took us to get down), God reminded me that this scenario wasn't so different from my own faith life. I often find myself at a standstill, fretting over something I think I need, when in reality God has already graciously gone before me and has everything I need waiting along the way. He simply asks me to walk down the stairs. And even though I've sung, "Though none go with me, I still will follow" countless times, am I willing to trust, to walk with him? Do I act on my belief that he has gone before and prepared the way?

May we learn to follow humbly and boldly after the God who is before all things and who holds all things together.

Discussion Questions

1. Are there places in your life where you need to remember that much has taken place prior to your arrival?

2. Are there situations in your life that you need to pray about, acknowledging to God, "You've got this"?

3. Where in your life might you act on your belief that God has gone before and prepared the way?

Notes

1. Wendell Berry, *The Art of the Commonplace: The Agrarian Essays of Wendell Berry* (Washington, DC: Shoemaker & Hoard, 2002) 20.

2. Maltbie D. Babcock, "This Is My Father's World," *The Baptist Hymnal* (Nashville: Convention Press, 1975) 155.

3. Eugene H. Peterson, *The Contemplative Pastor* (Grand Rapids MI: Eerdmans, 1989) 60–61.

4. Eugene H. Peterson, *Christ Plays in Ten Thousand Places* (Grand Rapids MI: Eerdmans, 2005) 2.

5. Colossians 1:18-20.

Chapter 6

Here and There

It is amazing what you can accomplish if you do not care who gets the credit.

—Harry S. Truman[1]

One of the most useful gifts I've ever received for my birthday was a bread machine. My mother- and father-in-law gave it to me a few years ago, and it is the only countertop jewelry that I own, always sitting on our counter in the kitchen. It's not a jewel tone and it doesn't sparkle, but it bakes hot bread and fills the house with wonderful scents, all at the touch of a button. This is an incredible invention. I mean, I can touch buttons on my computer all day long, and while it does what I tell it to do, my laptop doesn't give off any divine scents or warm the house as guests arrive. My husband, the electronics guru in the family, will disagree with me, but my technology of choice doesn't text photos or let me fling angry birds at pigs; it fills the air with a magical aroma and makes me a hot, flaky loaf that I can slice, melt some butter on, and drizzle with jam. In fact, I think I'll go pop a loaf in right now . . . you know, for research purposes.

That's the thing about a bread machine. As long as I keep the ingredients on hand, it takes me about five minutes to put them in the pan and press the buttons. Then, three or so hours later, I come back when the machine beeps and take out a golden brown loaf. No mixing, no watching the clock as it rises, no kneading, *nothing*. I don't have to understand how yeast works, something that always intimidates me in recipes. I don't have to understand all the different kinds of flours or what gluten is exactly. I plug it in, measure out a few ingredients, press a few buttons, and it's almost as if I actually know how to bake.

The bread machine has been a game-changer for the Sprinks. We now buy our wheatberries in bulk and go to Costco for fifty-pound bags of bread flour. With the number of sandwiches we make in our lunchbox stage of life, we're saving money and eating good food at the same time. But what was *really* exciting to discover was that if someone asks us over for dinner, I always have something people like that I can tie a ribbon around and bring to the meal. If I find out a neighbor is sick, it's not a big deal to bake a loaf in the bread machine and take it over. One year we gave loaves to friends as Christmas gifts, and people loved it. In a culture where food communicates love, I happened upon an easy way to care for people.

When I give bread as a gift or serve it at a meal, I can always tell who has never used a bread machine because they ooh and ahh over the "homemade bread" like I've been fretting over it in the kitchen for a whole day. "It's not homemade bread; it's bread-machine bread," I say. I'm quick to "fess up," even if others haven't seen the telltale paddle holes in the bottom of the loaf, because I am not much of a baker. Usually it takes just one bad loaf (which for some reason seems to happen when the weather changes) to humble me and remind me that I don't know what I'm doing. I'm just the recipient of a lovely gift.

How easy it would be, though, when we have people over to dinner, to sprinkle flour on my face and put an apron on as people walk in the door, greeted by the heavenly scent of fresh bread. It's tempting to feign the work of real homemade bread to people who don't know me so that I'm seen as a culinary genius. "Never mind that I recently called my mom to ask how to make corn on the cob," I think to myself, "I can float into the room and into the hearts of the people I'm hosting like a powdered-sugar-covered domestic goddess." It doesn't take long, though, for the real cook in me to show through, the one who once forgot to cut the wax off the hunk of cheese she bought and spent a good hour and a half stirring a rubbery cheese concoction that would *never* melt into sauce. I have to be honest. When it comes to bread, and cooking in general, I don't *have* a gift; I'm the recipient of one.

It occurs to me that I am tempted to do this in other areas of my life, to take credit for what has been generously given to me. My guess is that it happens easily to people who, on the whole, try to do the right thing in life. It's not even intentional most times. We don't purposefully feign goodness, but we go along in life, have a few successes, and then begin to believe that we are the embodiment of those wins. Then others comment on those wins, or on how neat we are, and we begin to drink in the belief that we are, without a doubt, some of the most fascinating people on the planet.

And what do we do? We use this position, this feigned idea of our own goodness, to our advantage. We use it to build ourselves up, sometimes even making further choices out of that persona. "Well, I could just serve *this* for dinner, but what would the *domestic goddess* serve for dinner? After all, I can't let others down," we think. We start operating out of such images of ourselves because it gives us confidence. When we begin to find our confidence and self-worth in these images, it's a dangerous place to be.

Paul talks about misplaced confidence as well, writing about the freedom he felt when he was released from having to live up to the righteous standards that he and other Pharisees had set for himself. You see, once he began following Jesus, Paul had to renounce the sin in his life and the way he was persecuting people who followed Christ, but he also had to renounce the good things about himself for which others would compliment him. Let's read what he writes about it in Philippians 3:7-9:

> The very credentials these people are waving around as something special, I'm tearing up and throwing out with the trash—along with everything else I used to take credit for. And why? Because of Christ. Yes, all the things I once thought were so important are gone from my life. Compared to the high privilege of knowing Christ Jesus as my Master, firsthand, everything I once thought I had going for me is insignificant—dog dung. I've dumped it all in the trash so that I could embrace Christ and be embraced by him. I didn't want some petty, inferior brand of righteousness that comes from keeping a list of rules when I could get the robust kind that comes from trusting Christ—*God's* righteousness. (*The Message*)

When he began to follow Christ, Paul not only had to renounce the bad things in his life. He also had to renounce the good things, things that pointed *only to Paul*, in order to fully credit Christ for the transformation that had taken place in his life. Morna D. Hooker puts it like this:

> To trust in something or someone means to rely on them, and complete trust suggests that there is no need to rely on anything else. So if men and women come to put their trust in God, they must abandon all other props. It is easy to think of faith in very positive terms, as acceptance—acceptance of the grace of God at work in Christ—and to forget this other, more negative aspect of faith—the need for renunciation. Before Paul could accept Christ, he had to renounce those things on which, as a Jew, he had relied (3:7-11). . . . Paul had to renounce the privileges that kept him from accepting the gifts that were now offered him in Christ.[2]

What does this mean on a practical level? It means that we need to come clean about who we are and how we got there. It means we need to tell the truth about who we are and *whose* we are. After all, we aren't fooling anyone. Christians who act like we have it all figured out, like we are who we are because of our own efforts, are like me, the bread machine button-pusher, deciding to open my own bakery. We're not gifted; we are the recipients of extravagant gifts from the ultimate Giver.

It's up to us to make sure that we don't take credit for the way God's grace and love have changed our lives. We need to make sure that others don't just think, "Wow, she's a great person," and put the credit where it isn't due. "The *real* believers," Paul writes in Philippians 3:3, "are the ones the Spirit of God leads to work away at this ministry, filling the air with Christ's praise as we do it. We couldn't carry this off by our own efforts, and we know it." The truth, if we listen to Paul *and* to our own hearts, is that anything and everything good about us is because of God's creative work in us and on our behalf.

This is a truth we need to keep before us constantly, lest we start to believe that we've had a little something to do with any goodness in our lives. The idea of continuous conversion is important here—not that we have to doubt our salvation but that we should live each day in awareness of our desperate need for Christ and his transforming love in our lives. The New International Version of Philippians 3:8-9 reads like this:

> What is more, I consider everything a loss compared to the surpassing greatness of knowing Christ Jesus my Lord, for whose sake I have lost all things. I consider them rubbish, that I may gain Christ and be found in him, *not having a righteousness of my own* that comes from the law, but that which is *through faith in Christ*—the righteousness that comes from God and is by faith.[3]

I think for most of my life I've read these ideas of Paul's and thought of him as a college student who decided to switch majors. His Pharisaic pedigree wasn't worth anything in his new emphasis of study, life in Christ, so he basically let his old coursework go as a sunk cost. But as I go on in life and realize my own tendency to forget how young I am in my faith, I realize that Paul was keeping before himself the reality that Christ is the cause of *anything* loving and right in his life. "Anything good in me," he says, "*is because of* Jesus' transforming work in me, because without it, I'm nothing." Paul is trying to warn me that any place where I am putting my confidence in other parts of myself is eventually going to lead me to the place where the gig is up and I come to grips with the reality that I'm not who I'm trying to be or who other people may think I am. Paul is trying to help us understand that our testimony is not just "My life was garbage before Christ." We must also be willing to tell the truth and say, "My life is garbage without Christ and it will be garbage in any place where I rely only on myself."

This is where we begin to examine the place toward which our lives are directing others. With every choice we make, we are ultimately saying either "Here," boosting our own image, or we are saying "There," directing others away from ourselves. With every conversation and every solitary action, we are either honest about who we are, or we are sprinkling flour on ourselves and pretending we can bake up a beautiful life on our own.

Let's be honest: we're all living off the incredible gifts God has given us. "We all live off his generous bounty, gift after gift after gift," says John as he begins to write his gospel.[4] Life, health, family, nature . . . the seemingly simple act of being able to drive to the grocery store and purchase food with money from your wallet is a grace from him. We are all existing, whether we take time to recognize our dependence on him or not, out of his generosity, and he *wants* us to. That's why he gives so many wonderful gifts to us—so that we will use and enjoy them. The only problem comes when I begin to stand on those gifts, buffeting the image I have of myself or judging others with them. The problem comes when I point to myself instead of to Christ, when I say "Here" when I should be saying "There."

If you were an artist interested in creating a work of art that focused on the biblical character of John the Baptist, you would probably need to study hands. You would do research by looking at all the paintings and mosaics created about him over time. And you would notice that John the Baptist is overwhelmingly portrayed with a finger pointing away from himself. You might even take a special class on how to sculpt or paint the human hand, how to create life-like knuckles and fingers and nails.

Artists or not, we can all see how John the Baptist has been portrayed throughout history as someone saying "There" instead of "Here." If you search the Internet for artwork featuring John the Baptist, you can see images that have inspired and mystified Christ-followers for years. Pastors and theologians such as Karl Barth have hung pictures of John the Baptist over their desks, reminding them of one who always shifted the focus away from himself. Yes, you can also see paintings of John baptizing Christ, and there are even some paintings where he is enjoying his infancy alongside Elizabeth, Mary, and the infant Jesus. But overwhelmingly, he is pictured as an adult with a finger or hand outstretched, beckoning us to look to Christ or images of the Lamb.

This is because John's understanding of his role in life, his primary ministry task as he knew it, was to say, "I am *not*," and instead point people to the "I AM." He spent his time helping people recognize and prepare for the things God was doing among them. Although people even followed him as disciples and copied his ascetic lifestyle, he worked hard to slip into the sidelines once he recognized Jesus had come. The Bible records him constantly saying, "Here he is! Here he is! The Lamb of God!" as Jesus walked by. He always took the spotlight off himself and put it where it was supposed to be, saying things like "He must become greater; I must become less."[5]

The Bible even tells us that devout people kept coming to John to ask him if he was the God-revealer, the Messiah, or not. If he had ever wanted to substantiate his austere desert lifestyle, to boost himself in the eyes of all the other religious people, here was his chance. But instead we read, "When Jews from Jerusalem sent a group of priests and officials to ask John who he was, he was completely honest. He didn't evade the question. He told the plain truth: 'I am not the Messiah.'"[6] Instead, we get a picture of a man "sent by God to point the way to the Life-Light. He came to show everyone where to look, who to believe in. John was not himself the Light; he was there to show the way to the Light."[7] We begin to understand the incredible way John was grounded in the reality of his place, a man whom many have artfully portrayed proclaiming, "There he is" instead of "Here I am."

Are we willing to do that? Are we willing to say, consistently, "There he is"? Are we willing to be completely honest, as John was, and say, "I don't know much about Jesus other than what God has revealed to me, but here . . . come know him for yourself. There he is."

John points to Jesus, calling him "the Lamb of God who takes away the sin of the world."[8] The form of the word *sin* in the original language is such that emphasizes not individual human sins but the world's collective brokenness.[9] John's work, as he saw it, was to direct people to the one who could fix his broken life and the broken lives of the people around him. He didn't use the changes God had worked in his own life to puff himself up. He didn't take credit for who he was, not even for the good in his life, but willingly pointed back to the one who changed everything so that they, too, could be changed. Will we be as willing, in following Christ, to point the way to him?

Recently, a dear friend taught me the super-helpful phrase "elevator pitch." As you may already know, this is the act of thinking about your words ahead of time so that when you are casually asked about a particular topic that is important to you—say, on an elevator—you have a concise answer ready to say that rightly expresses your thoughts. And as I meditate on John the Baptist, it occurs to me that I need to have an elevator pitch ready in this area of my life. Just as quickly as I can say, "It's not really homemade bread, it's bread machine bread," I need to be able to respond rightly and concisely to those who might comment about me. I need to be ready, as John was, to say, "I am not," and instead point to the "I AM." I need to be ready, as Paul was, to say, "We'll just give God the credit for that, won't we? Because everything I do on my own is pretty much garbage." I need to 'fess up if someone gives me a compliment. Somehow, I need to refrain from preaching a sermon and instead help that person understand that they're not seeing me; they're seeing a change God has worked in my life.

There's one more part of the human body that an artist would have to study in order to paint John the Baptist: the head. This is because the second-most way he's been portrayed in art over time is in the aftermath of his beheading. Others may understand the changes God works in our lives, or, as in John's case, they may not. The reality is that we don't let others determine our place, and we don't try to manipulate it ourselves. As John found out, one day people may respect and revere our message, and the next day they may shun us or even punish us for it. Instead, our obedience lies in our awareness of our place, our willingness to point away from ourselves, saying not "Here I am" but "There he is." May we who follow Christ live with an accurate and humble understanding of our place, always pointing to the giver of all the good gifts we've received.

Discussion Questions

1. Is there anything you need to tell others so that you can say not "Here" but "There"?

2. What is your elevator pitch when someone compliments you? How do you acknowledge God's work in your life?

Notes

1. http://thinkexist.com/quotation/it_is_amazing_what_you_can_accomplish_if_you_do/15345.html, accessed 5.18.12.

2. Morna D. Hooker, *Philippians*, The New Interpreter's Bible, vol. 11 (Nashville: Abingdon Press, 2000) 530–31.

3. Italics mine.

4. John 1:16.

5. John 3:30, NIV.

6. John 1:19, *The Message*.

7. John 1:6-8.

8. John 1:29, NIV.

9. Gail R. O'Day, *John*, The New Interpreter's Bible, vol. 9 (Nashville: Abingdon Press, 1995) 528.

Chapter 7

The Responsive Place

My imagination, however, requires that I stay in the same city, on the same street, in the same house, gazing at the same view. Istanbul's fate is my fate. I am attached to this city because it has made me who I am.

—Orhan Pamuk[1]

Being in relationship with people means responding to them. It means being available to them, asking and answering questions, and returning their calls. This is not a newsflash. We all say we want deep relationships with others, and that strong desire makes sense: we are created for and fulfilled by meaningful interactions with God and others. Connections of significance are some of the most beautiful aspects of our lives; they're also probably the parts of our lives at which we work the hardest.

At least, we hope that's the case. It is easy, especially when we find ourselves surrounded by the same people and places each day, to slack off. We find that place can quickly turn into space. The independent nature of our culture and the ways we over-commit ourselves often tease us into becoming people who are wholly unavailable to others. For example, when was the last time you asked people to RSVP to a party you were holding? How many people actually responded? We forget that being in relationship with others means responding to them. It means filling in that precious open space on your calendar with time for a real person.

I often find it funny how movies idealize this sort of thing. We want endless choices and possibilities with as little commitment as possible, and you can see this desire in our movies. Unless the movie is specifically about family, extended family isn't in the picture until they are needed, at which time they are extremely close. Another movie might show a person dealing with some sort of personal drama, which is alleviated when he returns to the never-before-mentioned home of his childhood, where he is suddenly greeted affectionately by childhood friends who seem to have been longing for his return. Movies are our escape. They give us ideal relationships minus the work.

We know in our hearts that the deepest relationships we have, the friendships we would categorize as life changing, involve our responses. They involve us making ourselves available when we are in the middle of other activities. They involve picking up the phone and responding.

And what happens when you respond? There is opportunity and potential. When you answer a phone call from a friend, you don't know what kind of story you will hear: a hilarious tale, bad news, or maybe encouragement. You might get a request to fulfill a need. You might feel guilty or angry because of something the person on the other end says. It might be the call you've waited for your whole life.

Sometimes, do we choose not to answer? Amid our busy, repetitive lives, do we choose to be unavailable to others around us? Do we choose to be unavailable to God? This is not a discussion about ignoring phone calls and texts at the dinner table: when we do that, we are choosing to be available to those with whom we're sharing the meal. I am talking about the unresponsive place where I sometimes find myself. It's actually good when I notice myself in this place; what's worse is when meaningful opportunities for love present themselves and I breeze by unaware. The truth is that our lives are not two-hour movies. Try as we might, we don't get to write the story of our lives. The best days are when I act on the knowledge that I'm *not* the author.

Recently in my reading, a verse from Isaiah 66 jumped out at me. God's people are trying to rebuild God's place, the temple. During a renovation that could outdo any HGTV episode, God says he's not concerned with the glamour of a place or the extravagance of a task. In fact, he tells them (and us) in verse 2 what he really wants. "There *is* something I'm looking for," God says, "a person simple and plain, reverently responsive to what I say."

He wants us to respond to him. If the phone rings, he wants us to shut the laptop and answer it. Here again is opportunity and potential. We don't know what we will hear: It could be a story that breaks our hearts. It could be a request or a need that we have the ability to fulfill. Have we made ourselves available to answer? If so, how will we respond?

God is looking for the reverently responsive. What does that mean? I don't think it means that he wants us to do his bidding mindlessly; he could have created us like that in the first place if that were his desire. Instead, I think it means God is joyful when I respond to him from a reverently responsive place without qualifying or dissecting his request. It means that I trust him and respond because I am in relationship to him.

There are times that God asks us to step up; there are times when the sin in the world demands it. The comforting reality is that God is always at work in the world. Most days, my role in that work might not find me in a Wonder Woman cape and boots ridding the world of injustice. More often it finds me in my flannel sheep pajamas emptying the dishwasher. And the question the Bible puts to me is, either way, will I respond?

Many times, I'm not reverently responsive to God *or* to other people. I'm prideful. I only want to respond if I get to wear the boots and cape and have people marvel at me. And if I don't get to be the superhero or if people don't notice my obvious Wonder-Womanish qualities, I decide against it. My pride is wounded and my insecurities get the best of me, which is the opposite of being reverently responsive.

Am I writing this to say that we must say yes to everything and everyone at all costs? No. I am writing against the spirit in me and, I imagine, the spirit in every man and woman that wants to work for good in the world, run home to see ourselves on the 10 p.m. news, and then spend another hour blogging about it. I am writing because many times, if situations aren't exactly what we thought, we make ourselves unavailable. We screen our phone calls, and if the need at the other end isn't to our liking, we don't pick up.

Being "reverently responsive" means I need to trust God and be attentive to the place where he has put me. I need not worry about the future or rehearse the past in my head. Norman Wirzba writes, "The place of health and happiness must be grounded in the place where we now are. . . . The test of our lives is not the past or the future. It is the present."[2] Our faith and trust in God, then, is exercised most consistently in the things God or others ask us to do *right where we are.* Of course, in reality, these acts of love *are* the most important thing. At holidays and times when we can look back at our lives over a warm cup of coffee, we know this to be true. But do we forget as we live out the details of life?

All through its pages, the Bible tells the story of finding meaningful relationships with God and others. It's filled with adventure and drama, and we read again and again how that adventure overtakes those who were reverently responsive. Many accounts don't even record the names of those who humbly made a difference wherever God placed them. For instance, consider the faith of a widow who came to Elisha, a prophet who was helping to make things right in Israel during the time of the divided monarchy. Her story is in 2 Kings 4:1-7, or in *The Message* as follows:

> One day the wife of a man from the guild of prophets called out to Elisha, "Your servant my husband is dead. You well know what a good man he was, devoted to GOD. And now the man to whom he was in debt is on his way to collect by taking my two children as slaves."
>
> Elisha said, "I wonder how I can be of help. Tell me, what do you have in your house?"
>
> "Nothing," she said. "Well, I do have a little oil."
>
> "Here's what you do," said Elisha. "Go up and down the street and borrow jugs and bowls from all your neighbors. And not just a few—all you can get. Then come home and lock the door behind you, you and your sons. Pour oil into each container; when each is full, set it aside."
>
> She did what he said. She locked the door behind her and her sons; as they brought the containers to her, she filled them. When all the jugs and bowls were full, she said to one of her sons, "Another jug, please."
>
> He said, "That's it. There are no more jugs." Then the oil stopped.
>
> She went and told the story to the man of God. He said, "Go sell the oil and make good on your debts. Live, both you and your sons, on what's left."

We don't know the name of this woman, the size of the jars, or the amount she received for the oil. But we know enough of the story to understand that even the simple act of borrowing jars, jugs, and bowls was a reverent response. Are we willing to do the simple things we are asked to do each day—and do them in faith?

If we read a little further into 2 Kings 5, we find another nameless person whose belief made a difference in the life of another. This time, it's a young girl who was captured from Israel during a battle and is made to work as a slave in the house of Naaman, the commander of the enemy army. Taken from her family, country, and everything she knew, she is forced into a life of slavery . . . and yet her faith emerges. As she discovers that her master, Naaman, suffers from leprosy, she says to Naaman's wife, "Oh, if only my master could meet the prophet of Samaria, he would be healed of his skin disease."

Hearing this and willing to try anything, Naaman travels to Elisha's doorstep deep within enemy territory with a grand show of horses, chariots, and gifts to give the prophet. The pageantry is interrupted, however, as Elisha sends a servant with only a message: "Go over there to the Jordan

River and take seven baths. Your skin will be healed and you'll be as good as new." Naaman loses it. He wants the fanfare, the drama, the superhero healing. POW! He is annoyed at the task because it is as commonplace as cleaning oneself. Moreover, he can't imagine lowering himself into the Jordan and would rather go back home to bathe in his own country's rivers.

And though he's asked to do something simple, right where he is, Naaman rages. He dissects the requirements. Take a bath to heal himself? Ridiculous! His unwillingness to do the simple task right in front of him sharply contrasts with the belief of the displaced young girl. Naaman's servants are incredulous and go running after him. "Father," they say to their master, "if the prophet had asked you to do something hard and heroic, wouldn't you have done it? So why not this simple 'wash and be clean'?" Naaman swallows his pride and consents. He bathes in the river seven times and is instantly healed.

In the same way he helped the widow and her sons toward life, Elisha helps Naaman toward physical and spiritual life. Naaman returns to Elisha (who doesn't send his servant but now comes to Naaman himself) and says, "I now know beyond a shadow of a doubt that there is no God anywhere on earth other than the God of Israel." When seemingly everyday things are done from a place of responsive obedience, life is renewed and those who are far from God are brought near.

Jesus reaffirms this truth to those in his hometown who were looking for the flash and dazzle. He says in Luke 4:21, "You've just heard scripture make history. It came true just now in this place." The people of his hometown are skeptical: how could someone as simple as a carpenter's son come and change everything? And then in 4:27, Jesus reminds them of Naaman: "There were many lepers in Israel at the time of the prophet Elisha, but the only one cleansed was Naaman the Syrian." He warns them not to be so proud that they miss the miracle of healing that could take place in their own lives. Unfortunately, we read in Mark 6:5 that they do miss out, for he "could not do as many miracles there." Their unwillingness to give themselves to the simple truth resulted in less wholeness in that particular place. "This is the way God cleanses people of their afflictions, it seems," writes Choon-Leong Seow, "not through the dramatic performance of a human healer, but through a simple act of obedience. Salvation comes mysteriously when we submit to God's script and not our own."[3]

We often try to turn our theology into autobiography by dreaming up our own story. Instead, what if we noticed that the author of our biography has already set a grand piece of the story in motion? He has determined the setting of our story and asked us to immerse ourselves in it in reverent responsiveness. It isn't so much about *doing* as it is about a lifestyle of active *being*.

I am working at responding to God and to others reverently and in a timely manner. I'm working at being available as I go through my day, especially when my attention is needed at a time when I didn't necessarily plan to give it. When you have children, you realize that you can't schedule or orchestrate times of connection; when they are ready to open up or when they need snuggle time, you'd better put down what you are doing and be with them. Otherwise, you miss it.

It's the same with our relationships with God and others: you can try to plan times of connection, but more often you must be willing to connect with others when they need you, no matter what you've already planned. The author of the book written to the Hebrews agrees: "God takes particular pleasure in acts of worship—a different kind of 'sacrifice'—that take place in the kitchen and workplace and on the streets" (13:16).

In our availability, in our trust of God and the place in which he's put us, we are on the lookout. We are not resigned and not consumed; instead, we're watching for the moments where we need to lean further in faith. That quiet act of leaning in may make a difference to us or to another; it may be the reason God put us here in the first place. "There is treasure buried in the field of every one of our days, even the bleakest or dullest," says Frederick Buechner, "and it is our business, as we journey, to keep our eyes peeled for it."[4]

Maybe we need to retire our capes and boots and look to the surprising, mysterious author of this story. Maybe we need to stop trying to figure out God's will for our lives and look at the context clues he's already given us. It may be that our wholeness lies in responding reverently to God and to others—and responding with knowledge that the end of the story is already written.

Discussion Questions

1. To whom or what might you need to respond in your place today?

2. Do you see responding to others as an annoyance or as an act of worship as described in Hebrews 13:16?

Notes

1. Orhan Pamuk, *Istanbul* (New York: Alfred A. Knopf, 2006) 6.

2. Norman Wirzba, "Introduction," in Wendell Berry, *The Art of the Commonplace* (Washington, D.C.: Shoemaker & Hoard, 2002) xviii-xix.

3. Choon-Leong Seow, *2 Kings*, The New Interpreter's Bible, vol. 3 (Nashville: Abingdon Press, 1999) 198.

4. Frederick Buechner, *The Longing for Home: Recollections and Reflections* (San Francisco: Harper, 1996) 120.

Chapter 8

Innermost Places

Always forgive your enemies; nothing annoys them so much.
—Oscar Wilde[1]

I was recently reading the news online and came across an article that surprised me. The headline caught my eye and I had to click on it: "Blaming Others Can Ruin Your Health." The main idea was that people living in a continuous state of resentment toward others directly affect their physical bodies in profound ways. The body's hormone and immune systems are affected by emotions of bitterness, which results in higher blood pressure, increased heart rate, and a higher likelihood of heart disease. The trend has caused some in the medical and psychological field to propose a new diagnosis called PTED, post-traumatic embitterment disorder to categorize people who can't forgive what others have done to them. To me, however, this was the kicker: "The data that negative mental states cause heart problems is just stupendous," said Dr. Charles Raison, of Emory University School of Medicine. "The data is just as established as smoking, and *the size of the effect is the same.*"[2]

What? I thought. While many health-conscious Americans wouldn't make a habit of smoking, they are unknowingly taking drags of a substance just as toxic: resentment. At first, it might not seem like there's an issue. We might express our frustrations socially to gain the empathy or friendship of others. What's the harm, we say, in gathering on the stoop outside our offices and taking gripe breaks? It doesn't control us, we think. After all, *we're* in the right. Someone else has done wrong to *us*.

What we don't realize, however, is that bitterness, too, is addictive. According to this article, it follows us, body and spirit. Others begin to detect its scent as we enter a room. They notice the aroma on our clothes and belongings. Before long, there are secondhand effects; our homes reek of our resentments, and our children take them on early as well. Eventually we discover that our habit of bitterness has polluted our entire lives. It permeates us inside and out.

And, while it's not printed on the outside of a carton, we have been warned of resentment's toxicity. "The wages of sin," says Romans 6:23 (NIV), "is death." Parts of us die when we give in to sin. We're not merely crossing a boundary; we're diminishing our capacity to function in the way God intended. Sin pollutes and deteriorates us, no matter if it is of our own doing or that of another. "The great tragedy of life is not death," said writer and citizen-diplomat Norman Cousins, "but what dies inside us while we live."[3]

Fortunately, God is in the business of resurrection. Science, it seems, only confirms Scripture in our need for detoxification. We must rid ourselves of what diminishes us and allow space for love and truth, especially in our innermost being. "Surely you desire truth in the inner parts," writes the psalmist. "You teach me wisdom in the inmost place" (Ps 51:6, NIV). In other words, we must not let our sins or those of others take up valuable space inside ourselves. We must not constrict our capacity to love.

When I talked about these ideas with my husband, he told me that the most poignant message he remembers from a health class was the day his professor asked everyone to meet in the stairwell of the building. Then she handed them each a straw and asked them to begin climbing to the top of the stairs, breathing in and out only through the straw. Smoking, she said, decreases your lung capacity in such a fashion. It limits the amount of fresh air you can take in.

Is this like us when it comes to our resentment? We know we shouldn't hang on to it, that it's not the healthiest thing for us, and yet many times we choose to do so. The tragedy is the big picture: we can only draw breath through the straw. We ultimately diminish our capacity for love. This inhibits our ability to draw in God's life-giving elements and limits the energy we have for others. Why do we give our frustrations that space in our lives? Why do we give our sin and the wrongs others have done so much power over us? We must not be controlled by our sins or those of others. We must get rid of the pollution, the litter, in our innermost places.

We sing a song in my church that contains the phrase, "Sin has no hold on me." It is always a compelling moment during worship when people join together and sing that phrase repeatedly. It is as if the whole group realizes that Jesus died to rid us of the polluting effects of sin and allow us to function without its limitations. As the reality of what Christ has done for those who follow him washes over us, we discard the straw and breathe life in deeply, the way he meant for it to be lived. For sin, ultimately, has no hold on us.

What does "discarding the straw" look like? What does it mean to remove the issues that constrict us? Is there a method? Is there a patch for our resentments? Can we quit cold turkey? While the process might vary depending on the person, we see in the Bible that confession and forgiveness are two central practices in a life of following Christ. The writers of Scripture urge us, for our own good, to continuously embrace the detoxifying effects of confession and forgiveness.

Let's forget about others and begin with ourselves. It is probably safe to say that confession of our own mistakes is more time-consuming, because while we can quickly recall those of others, we bury the ones we commit deep down. We don't like full disclosure. But we see in Scripture that God sets up confession as an early practice for those who follow him.[4] In order for his people to live fully and not diminish the capacity he created within them, God prescribes frequent doses of confession for humanity. Our lists of sin do not placate him, nor does he care to hear emotionless accountings of our issues. Rather, we need time, it seems, to slow down and let our actions bother us a bit, as it says in Psalm 38:18 (NIV), "I confess my iniquity; I am troubled by my sin."

When it comes to sin, we don't shake it off or tough it out, hoping nobody notices. "You can't whitewash your sins and get by with it; you find mercy by admitting and leaving them" (Prov 28:13). And though it's against everything our culture prizes, God asks us to have some kind of continuous process, some kind of ongoing mechanism, of confession in our lives.

I say ongoing because we sometimes forget that since many of us have, at some point in our lives, recognized and confessed that we are sinners, we live as if that one-time event has put us into a lifelong state of humble contrition. We confess our sins, receive salvation, and believe we're set right with God . . . and so we are. We all must agree, however, that experience confirms how quickly we move beyond our regret. All too quickly, we begin to forget the grace we've received. God doesn't want us to linger in guilt, but, as Richard Foster rightly reminds us, "The Bible views salvation as both an event and a process."[5] We must not be so confident in our salvation that we forget to allow our own sin to grieve us; instead, we must be confident in our God, who allows us to come to him and confess continuously, that we may be restored.

Many times, confession is a private process that occurs between God and us, his children. At other times, it is a communal event, as in James 5:16, where we read of a pastor asking a church to "confess your sins to each other." I enjoy praying the prayer below, part of Morning Prayers from an Anglican prayer book I was given. When the faith community (or just two members of it) is gathered, everyone reads the **bold** words together.

> Let us call to mind and confess our sins.
> (silence)
> **Almighty God, our heavenly Father**
> **in penitence we confess**
> **that we have sinned against you**
> **through our own fault**
> **in thought, word, and deed**
> **and in what we have left undone.**
> **For the sake of your Son, Christ our Lord**
> **forgive us all that is past**
> **and grant that we may serve you**
> **in newness of life**
> **to the glory of your Name.**
>
> Almighty God have mercy on us, forgive us our sins and keep us in eternal life; through Jesus Christ our Lord.
> **Amen.**[6]

No matter the form, confession is something that God, who created us, tells us that we need—and need *continuously*. It is to our detriment when we choose to breathe through the straw and limit God's power in our lives. Instead, we can willingly and repeatedly come to God, who clears the pollution out of our innermost places.

A second practice we need to exercise is that of forgiveness. We all say we need it and desire it from God and from others. We might even choose it when we find ourselves in a weighty relational predicament. Something we see in Scripture, however, is that forgiveness is also an ongoing practice, a continuous characteristic of the life to which we are called. When Jesus was teaching us how to pray, he said we should say to God, "Keep us forgiven with you and forgiving others."[7] He knew this because he was communicating to us the nature of God, who consistently forgives. "If you, GOD, kept records on wrongdoings, who would stand a chance?" writes the psalmist.

"As it turns out, forgiveness is your habit, and that's why you're worshiped" (Ps 130:3-4).

What practical difference does this make? It means that instead of choosing to forgive someone only when you feel that the airplane is going down, you make a practice of offering forgiveness every time you feel a little turbulence. Forgiveness isn't something we piously offer only when an emotional blow-up is followed by a request for restoration. It may, in fact, only require you. No matter if restoration is requested or not, forgivness is something we must boldly and quickly choose. It is a lifestyle. "Life is an adventure in forgiveness," writes Norman Cousins.[8]

Adventure, we know, involves risk. When Christ was leading us to forgive and showing us how to do it, he wasn't ignoring the facts or living in an idealistic bubble. Instead, taking all realities into account, he audaciously chose to forgive, again and again. It didn't matter to him what others thought; he didn't try to maintain the relational upper hand. In this way, probably one of the most courageous moments we read in all of Scripture is not the story of David and Goliath or Daniel in the lion's den. It is the simple prayer, "Father, forgive them," that Jesus said while hanging from the cross. As followers of Christ, practicing continuous forgiveness may be the most courageous way we can live.

Nelson Mandela knew the risks of forgiveness. As the newly elected president of a South Africa fresh from struggle, forgiveness made him vulnerable. Clint Eastwood captures this risk wonderfully in a scene from the movie *Invictus*. A few minutes into the movie, we find the head of presidential security, Jason Tshabalala, in his office with the rest of his team, worrying that more men are needed to provide adequate security for the president. To his surprise, four Afrikaans officers walk into his office, having been assigned to presidential security by Nelson Mandela himself.

Tshabalala is incredulous. These men were the bodyguards for the *last* president; they worked for the government that suppressed, fought against, and tortured black South Africans like themselves. These Afrikaans men were the enemy, and now they were assigned to *protect* Mandela? He rages into the president's office for an explanation.

"Ah, yes," says Mandela. "These men are special trained . . . they have lots of experience. They protected de Klerk."

Tshabalala protests.

"You asked for more men, didn't you?" says Mandela. "Reconciliation starts here." He hands the orders back to Tshabalala, who still can't understand.

"Not long ago, these guys tried to kill us," he says angrily to Mandela. "Maybe even these four guys in my office . . . tried and often succeeded . . . ."

"Yes, I know," says Mandela, face to face with Tshabalala. "Forgiveness starts here, too. Forgiveness liberates the soul. It removes fear. That is why it is such a powerful weapon."[9]

Perhaps we should listen to these voices of one who, after twenty-seven years in a jail cell, was ready to forgive and of one who forgave as he hung on a cross. Perhaps there is freedom, spaciousness, that we can't yet comprehend. "Resentment," said Mandela, "is like drinking poison and then hoping it will kill your enemies."[10] These voices remind us that in holding on to our hurts, we poison ourselves.

It should not come as a surprise to us that we hoard our resentments. We live in a culture consumed by and dependent on tangible acquisition, and it is easy to follow this pattern in spirit. We might regularly go through our houses and tidy up, ridding ourselves of what we don't need, but all the while we are spiritual hoarders, holding on to this or that because it justifies us. As the earlier article mentioned, it is as unhealthy for your spirit to hoard the "pride" of bitterness as it is to live in a house full of garbage.

And the Bible says it is unnecessary. It's not how we were meant to live. Health awaits us. Life awaits us. In Isaiah 58:9-12, the prophet writes,

> If you get rid of unfair practices, quit blaming victims, quit gossiping about other people's sins, if you are generous with the hungry and start giving yourselves to the down and out, your lives will begin to glow in the darkness, your shadowed lives will be bathed in sunlight. I will always show you where to go. *I'll give you a full life in the emptiest of places*—firm muscles, strong bones. You'll be like a well-watered garden, a gurgling spring that never runs dry. *You'll use the old rubble of past lives to build anew, rebuild the foundations from out of your past.*[11]

Forgiveness, the prophet says, is what we *need*. It's not just a good idea or even merely our duty before God. It is what will give us the power to build anew amid the rubble of our lives. It is the power to boldly rebuild a warring nation. It is resurrection power.

Maybe we need to have a few spiritual garage sales. If we regularly rid our houses of our tangible junk, maybe we should also add to our calendars a regular time to simplify our lives spiritually. Confession and forgiveness require intention and time. We might need to rework schedules or set up reminders that will help us to "clean house." We may need to invite accountability or get away for a while. When she was away at school, author bell hooks (whose pen name is intentionally lowercased) noticed her need to clean house. In her book *Belonging: A Culture of Place*, she writes,

> Away from home I was able to look back at the world of homeplace differently separating all that I treasure, all that I needed to cherish, from all that I dreaded and wanted to see destroyed. Like a country estate sale where all belongings are brought from a private world and are publicly exposed for everybody to gaze at them, pick them over, choosing what to reject or keep, ultimately deciding what to give away or just dump, away from home I was able to lay bare the past and keep stored within me much that was soul nourishing. And I was able to let much unnecessary suffering and pain go.[12]

Sometimes, it is only in being willing to get out of our place intentionally that we can truly be at home in our innermost place. Sometimes, it is only in discarding the straw that we are able to inhale in the spacious life God has for us.

May we who follow Christ boldly go after him in the ongoing adventure of forgiveness.

Discussion Questions

1. Are there straws in your life that you need to discard so that you can breathe deeply?

2. Do you have an ongoing practice/mechanism for confession?

3. Is forgiveness your habit? Your adventure?

Notes

1. Oscar Wilde, http://www.quotationspage.com/quotes/Oscar_Wilde/ (accessed 8 November 2011).

2. Elizabeth Cohen, "Blaming Others Can Ruin Your Health," http://edition.cnn.com/2011/HEALTH/08/17/bitter.resentful.ep/index.html?&hpt=hp_c2 (accessed 18 August 2011), italics mine.

3. Norman Cousins, "Norman Cousins Talks on Positive Emotions and Health," Santa Monica, transcription from original radio broadcast, KCRW-FM, 1983, http://www.bobrosenbaum.com/transcripts/nctalks.pdf.

4. See Leviticus 5:5, for example.

5. Richard J. Foster, *Celebration of Discipline,* rev. ed. (San Francisco: Harper, 1988) 145.

6. *An Anglican Prayer Book* (South Africa: HarperCollins, 1989) 45.

7. Matthew 6:12.

8. http://www.quotationspage.com/quote/31318.html (accessed 8 November 2011).

9. "Forgiveness Starts Here," *Invictus,* DVD, directed by Clint Eastwood (Burbank: Warner Brothers, 2009).

10. http://www.goodreads.com/author/quotes/367338.Nelson_Mandela (accessed 8 November 2011).

11. Italics mine.

12. bell hooks, *Belonging: A Culture of Place* (New York: Routledge, 2009) 61.

Chapter 9

Potential in Locality

> *Biblical religion has a low tolerance for "great ideas" or "sublime truths" or "inspirational thoughts" apart from the people and places in which they occur.*[1]

I am a stay-at-home mom. I write and have other involvements, but the things I spend the most time doing daily are not rocket science: I'm simply helping the Sprink family to run smoothly. Each member in the family has his or her responsibilities, and we are learning, slowly but surely, to do them independently.

This morning, I was pleasantly surprised to find that my bed was the last one made. Progress at last! Bed-making was a struggle at our house: before each of them reached age four, my children told me on separate occasions that they should not be required to do such a thing because making the beds was *my* job.

Argh. That kind of statement doesn't go over so well with me. *Where do they get such an idea?* I wonder. Am I some person who exists just to make sure they have clean teeth, tidy hair, and organized rooms? Sure, I help them with those tough-to-reach corners when they are making up their beds. If they forget to make them, I give a little grace and make them myself. Nothing, however, squelches that spirit of grace more than hearing, "That's *your* job, Mom." Surely they're not purposefully leaving the bed *un*made because they know I will come along and make it for them?

They're children. They're still learning. Eventually they will grow up and realize that I am a person, too. They will, at some point, understand that I have goals and dreams of my own, none of which include making beds but most of which include *family*. Hopefully, children grow and move beyond their kid-centered view of the world.

Hopefully, as children of God, we do the same. Adults have growing up to do as well. It is our job, as followers of Christ, to move beyond a view of God that treats him like a stay-at-home dad who exists solely to make us look good. We are childish in our understanding of him if we think of him as a magical maid who is hanging around waiting for opportunities to pick up our junk. First and foremost, we must begin to understand that God is not a supernatural supporting actor in the story of our lives. We still have some growing up to do when we treat him this way. More important, there is so much more to him and to his idea of *family*.

The truth is, even though I want my kids to keep their rooms clean, I gave them their rooms for many other reasons. I want them to have their corners of the world, specific places where they can rest, play, imagine, and feel safe. I want them to put a little of themselves into their rooms—their personalities, their likes and dislikes. I want them to color their places with all their favorite colors (okay, maybe not literally, like with Sharpies, but you know what I mean).

I want them to have places where they can learn and study in peace and to pursue interests that lead them far beyond the walls of our home. I don't give them their rooms with the expectation that they should keep them looking like museums; I want them to *play* with their toys, to draw and pin their work on their bulletin boards. Ultimately, I want them to grow in their rooms. I want the spaces I've provided for them to be places where they can become *Lucy* and *Mikias*. Put simply, I've given their rooms to them out of love.

We, too, have been named family, and we find ourselves living and growing in a specific place. It is a place with potential and opportunity. It is a context for connection with others and with ourselves, a place where we can grow emotionally, physically, and spiritually. God wants who we are, and the creative spirit he put in each of us, to spill out into our place. Put simply, he gave us our places out of love, and he wants us to fully *live* in them.

As we grow into our places, we begin to mature and recognize that there are responsibilities that come with our place and exciting reasons for the tasks we've been given. Life in God's family is not about appearances or pretense. Just like I'm not asking my kids to clean their rooms only so they'll look good, God is not sprucing up his family for the purpose of looking holy to others. He isn't into creating a holiness museum-like world where we never get to enjoy and live into the things he's placed around us.

The gift of salvation, our very spiritual life, is not about God making a "better me," one that stays pristine and separate from others. Yes, when we begin to emulate Christ, when we accept his grace and let him change our lives, the fruit of the Holy Spirit begins to characterize who we are. Yes, we are changed and cleaned and reorganized, *but for what purpose?* There is so much more to the lives he's given us, so much potential embedded in the places we find ourselves and in the tasks he asks us to do. We, as God's children, would do well to look for it, to engage it, and to talk to him about it.

We must look for the context clues. For example, no matter what fun is up ahead, there is always resistance when I ask my kids to stop what they are doing and clean up what they are playing with. They're kids. It's not their fault. They still have some growing up to do. They live completely in the present. All they know is that there is something else they want to be doing *other* than the bit of work I'm asking them to do now.

The kids don't realize that having a fairly tidy house frees us to be able to have our friends over at a moment's notice. They don't understand that having a clutter-free living room floor allows us to plop down on it and play board games together—board games that still have all the pieces because we picked them up and put them away last time. The children can't see yet that there are other fun things up ahead on our family calendar, and that if we get some minutes of housework done now, we will be free to spend carefree hours together later.

Parents, on the other hand, see the big picture. I see what's up ahead: the bus is coming in half an hour . . . the spelling test is on Thursday . . . we're leaving town on Friday. I know our family's goals, and I know what it's going to take to get us there. I know the time involved in getting us out the door, and I know what needs to be completed beforehand. As a parent, I know that being a family is more than just sharing space; it requires timing, cooperation, and a common vision, even if that vision is simply having a little extra time to go out for frozen yogurt this week.

What we see as we move on in our understanding of God is that his family is like this. While we may be sitting in our places tinkering with our toys, there is so much more going on. Instead of pouting when God asks us to do a little work in the places he's lovingly given us, what if we children trusted in his purposes and his vision for our world? Would we see that when God asks things of us, he might simply be trying to free us from our messes so that we can get on with the extraordinary purposes he has in store? We must make a concerted effort to look for the context clues in our places. Let's be bold enough and mature enough to ask for a glimpse of the Father's plan for our lives and for the world, for surely our childlike faith will draw us closer to him.

Paul talks about the purpose God gives us in our particular places in his first letter to the Corinthian church. "Where you are right now is God's place for you," he says in 1 Corinthians 7:17. "Live and obey and love and believe right there." In this chapter, he speaks about husbands and wives, slaves and masters. Paul is not dictating specific ways of life here, nor is he endorsing slavery or marriages that look like slavery; he is using these situations to argue that *all* circumstances, *all* places, are within the reach of God's loving potential and that every life, every place, is capable of wholeness in his grace. "Location and setting are indifferent matters," writes scholar J. Paul Sampley. "One's call is not. The gospel can flourish and be walked in any circumstance, and the living of it elevates the person and the circumstance in which the person lives."[2]

Further, Paul's challenge to live with purpose in our specific places is not a call to inaction or the demand to stay permanently in a particular situation. After all, we don't give our children rooms of their own so that they can hide behind their closed doors. Paul's challenge is a call to work toward and uphold the kingdom values of the family in the places in which we find ourselves. "We ought to appreciate and celebrate more fully that it is not just individuals who are changed for the better by their living out the gospel," writes Sampley. "Their circumstances may also be elevated by the gospel's power; their context and those persons around them may be positively affected by the reverberations of God's grace in and through them."[3]

So we see that we have been lovingly given a place in this world and that there is potential in it. We see that we have a responsibility to keep up our places and not hole away inside them. And the question I have to ask myself each day is, does my place look better or worse because I am living in it? Am I upholding God's rightness in my place or not? Am I a part of the problem or a part of the solution in Blue Springs, Missouri?

It begins with me growing up a little, with realizing that my life, just like the lives of my children, isn't about doing or saying whatever I want whenever I want. Recently, I was writing in a coffee shop and noticed a customer walk up to the counter. She waited there for an employee to greet her and then said in a harsh voice, "I just want you to know that *that* was the absolute *worst* scone I have ever had." Then, the lady turned on her heel and walked away, not giving the employee behind the counter a chance to say anything.

Really? I thought. Out of all the wrongs in the world to get mad about, we are now choosing to get huffy over *scones*? It was an obvious attempt not to provide feedback but to offend. I'd be surprised if the employee behind the counter even made the scone in the first place, yet I imagine that exchange darkened her day.

The last sentence of the book of Judges (21:25) talks about this. It points out the cause behind the absolute chaos and pervasive disregard for others that the whole book discusses: "People did whatever they felt like doing." If we all, like the coffee shop lady, choose to do what is right in our own eyes whenever we want, our cities will suffer for it.

Paul reminds us that we all began life with this childish, me-centered perspective. In Ephesians 2:1-5, he writes,

> It wasn't so long ago that you were mired in that old stagnant life of sin. You let the world, which doesn't know the first thing about living, tell you how to live. You filled your lungs with polluted unbelief, and then exhaled disobedience. We all did it, all of us doing what we felt like doing, when we felt like doing it, all of us in the same boat. It's a wonder God didn't lose his temper and do away with the whole lot of us. Instead, immense in mercy and with an incredible love, he embraced us. He took our sin-dead lives and made us alive in Christ.

God is gently and patiently helping all of us learn how to keep up our places, not so we can brag about having a pristine room of our own but so the rightness and wholeness of his family will permeate the entire world. It is God's gift and not our own doing, says Paul. "We are God's workmanship, created in Christ Jesus to do good works, which God prepared in advance for us to do."[4]

We need to be aware of our context as individual believers and also join together with others in our particular places to enjoy and extend life in Christ. God did not give us our places so that we could sit in our rooms behind closed doors. This new life we have in Christ is not just about salvation, and it's not about making sure others' rooms are clean as well; instead, we are to seek and work toward Christ's wholeness in all its forms.

The incredibly exciting thing is that you and your church are the *perfect* ones to do this. You have, as it says in Ephesians, been created to extend Christ's wholeness and healing to the people in your specific context. The church is "that part of the society that has a new relationship to God yet reacts in terms of the attitudes and presuppositions of that society."[5]

What that means is that you, because you know your place and the people in it so well, have a better chance of connecting with and understanding those around you. Why? Because you think like them and act like them . . . because you *are* one of them. There are no cross-cultural obstacles for you in your place. You are already equipped to make a positive difference in your place because it's, well, *yours*. You know its basketball teams, its biggest employers, and how the community felt the last time tragedy struck. You know your place; you probably look and act and talk like your place. You don't even need to think about it—your place is just there, a part of you.

That's why you and your church are perfectly equipped by God to reach out and make a difference in your place. In *Missional Church*, Darrell Guder writes, "The church's essence is always embodied in some tangible, visible form that is shaped by its particular time in history and its place in some specific human society."[6] We blow it, though, when we begin to think that our room is the epitome of all rooms. If we begin to exclude others by locking the doors to our rooms, playing only our favorite music, and focusing only on our favorite truths, we miss the point of having our own places. We miss the meaning of being *family*.

"The church has tended to separate the news of the reign of God from God's provision for humanity's salvation," says Guder. "This separation has made salvation a private event by dividing 'my personal salvation' from the advent of God's healing reign all over the world."[7] What does this mean? It means that you and other Christ followers in your place should work for change where you are. It means that instead of trying to be a moral room monitor in your community, you should model and interpret the freedom and spaciousness of life that comes from knowing Christ. As Guder writes, "The particular mission community is always involved in the discipline of becoming culturally bilingual, learning the language of faith and how to translate its story into the language of its context, so that others may be drawn to become followers of Jesus."[8]

Are we involved in change as followers of Christ? Or are we just playing in our own rooms? Are we as the church on the front lines of change in our places? Are we the first to battle the injustices in our communities? Why shouldn't we be? Are we asking questions of ourselves and engaging the community around us? Are we sensitive to the ways God has gifted and enabled our churches to meet the felt needs of the communities in which we live? Or are we satisfied living in our own Christian messes, not willing to do the work involved in keeping our places clean?

Once, while the four of us were driving in the Sprink station wagon, Lucy noticed Matt (who was driving) wave to a man walking on the side of the road. "Who was that, Daddy?" she asked as the light turned green and we kept going. "That's a friend of mine named Fred," Matt answered. He said no more to us but looked into the rearview mirror and said under his breath, "He's wearing my shoes." Later, I asked him about it and he told me Fred had come to the church a few days earlier and asked for some clothes, specifically shoes. Matt and Fred found they shared the same shoe size, and, since our church didn't have a clothes closet, Matt simply gave him a pair of his.

I tell that story not to cheer for Matt but to cheer for a God who equips us to make our cities better. Shouldn't we long to drive around the places in which we find ourselves and *literally see the changes God has worked*? If we trust God's direction in our lives, we stop doing whatever we want and instead pass on the graces we've received so that our contexts look different. "Be generous with the different things God gave you, passing them around so all get in on it," writes Paul in Ephesians 4:10-11. "If words, let it be God's words; if help, let it be God's hearty help. That way, God's bright presence will be evident in everything through Jesus, and *he'll* get all the credit as the One mighty in everything—encores to the end of time. Oh, yes!"

The grace of it all is that our job is more about trust than responsibility. The truth is that we can't do it all on our own, and we can quickly get overwhelmed if we try. There are corners of our beds we can't reach, and we are ill equipped to do all the right-making work ourselves. Jesus said, "He knows better than you what you need. With a God like this loving you, you can pray very simply. Like this: Our Father in heaven, reveal who you are. Set the world right" (Matt 6:8-10). It's a grace we dare not presume upon but one in which we can trust, knowing that God's goal is to connect us all as true *family*.

Discussion Questions

1. How are you looking for the context clues in your life? Are you talking to God about them and engaging them?

2. Does your place look better or worse because you are living in it?

3. How are you doing at becoming culturally bilingual (learning the language of faith and translating its story into the language of your context)?

Notes

1. Eugene H. Peterson, "Introduction to Joshua," *The Message* (Colorado Springs: NavPress, 2003) 232.

2. J. Paul Sampley, *1 Corinthians*, The New Interpreter's Bible, vol. 10 (Nashville: Abingdon Press, 2002) 880.

3. Ibid., 883–84.

4. Ephesians 2:10, NIV.

5. Dale W. Kietzman and William A. Smalley, "The Missionary's Role in Culture Change," in *Pathlight: Toward Global Awareness*, ed. Meg Crossman (Seattle: YWAM Publishing, 2008) 105.

6. Darrell L. Guder, ed., *Missional Church* (Grand Rapids MI: Eerdmans, 1998) 86.

7. Ibid., 92.

8. Ibid., 237.

Chapter 10

Panoramic View

> *Wise men and women are always learning, always listening for fresh insights.*
>
> —Proverbs 18:15

Once my husband called mid-morning to see how my day was going. It had been one of those days, a long day already, even though it had just begun. The details of my work and the challenges of childrearing made me wish it were already bedtime. I wanted to go to sleep and start over, and I probably said something to that effect to Matt on the phone. "Let's go to lunch," he said. "Let me take you out."

He was right. I needed to reboot. Unfortunately, it was not in the cards. The lunch order at one of our favorite Asian places was botched, and my preschooler's mealtime antics just added to my irritation. Tired, I made it to the end of the meal and cracked open my fortune cookie only to read, "Your life is a dashing and bold adventure."

Ha, I laughed to myself. Probably the boldest thing I'd done that day was to buy soymilk instead of skim, and the only dashing I did was across the restaurant like a Heisman Trophy winner with a kid saying, "I need to go potty, Mommy!" tucked under my arm. Who writes these cookie fortunes, anyway? Adventure? All I wanted to do was complete my duties for the day and crawl back under the covers. Adventure was too much to ask for that day.

Sometimes I wonder about that attitude in me, especially with regard to my faith. I say I want adventure, and I tell God that in my prayers. I probably even sing about it in worship, but *do I really want adventure?* Am I so tied up in doing my duties, my life stuff, that I laugh at the idea of change? Do I live a boring life, or does the dashing and bold life actually come knocking? Have I been so concerned with maintaining the faith life I know that I say "Pass" when a new opportunity for faith comes along?

Our view of our place is limited. We tend to live out of our experiences and history, which is a good thing. We're focused on what is directly in front of us, just trying to take the next steps. But our view of the world, influenced by locality is just a partial view. If we look up, we see an incredible panoramic landscape of the world that God wants to show us. This endless, scenic vista requires only our commitment to go to him continuously, asking where to look next. If I'm not following him, I might only see what is right in front of me.

The more I read my Bible, the more I find I can't be so judgmental of the Pharisees in Jesus' day. It was easy early in my faith to roll my eyes at them, to read the Gospels and shake my head at the fact that they weren't "getting it" when all the while Jesus was right there in front of them. Because they so frequently challenged Jesus, I had always written them off as the "unbelievers," the ones who didn't want anything to do with God.

What we see in the Bible, though, is that the Pharisees weren't the ones who didn't believe in God; they just didn't believe that God could work in the world through someone like Jesus. They weren't the "bad guys"; they loved God and wanted others to love God, too. They loved God by being strict observers of the Mosaic traditions, the moral codes that helped maintain their faith. They figured that the best way their minority group could regain power in the region was to engage in codified, uniform expressions of faith (Acts 15:5; 26:5). Maintenance of these religious acts was key, and any ripple in that strategy was not tolerated since it would hinder their campaign for political power.

And so they missed it, or, rather, they said, "We pass." Here and there, we read of a Pharisee who was willing to risk it, to hope that God was behind the right-making actions of Jesus, but for the most part, the Pharisees decided the adventure was just outside their understanding of God. They concentrated on their own, limited view of how God works and dismissed the possibility of the panoramic view.

There were others, though, who were willing to lean in further. There were twelve who were named and others who followed anonymously. Some were new to this whole Yahweh thing, but others had been looking and longing for God to show up in their everyday lives. John 1 tells of two men, one of them named Andrew, who sought God by doing religious acts like participating in baptisms. They were watching John the Baptist, and one day they heard him point to Jesus and say, "Here he is, God's Passover lamb." They decided to follow Jesus. They hung out with him for the day. They brought their friends and relatives to meet him. Then they left their fishing nets and went on the greatest faith journey of their lives, one to which they would ultimately give their lives.

They explored who Jesus was. They saw Jesus heal sickness, win spiritual battles, and command nature. They heard him laugh, learned what made him cry, and watched him love people who never got loved. They learned about who God was and how he loved humans. They learned truth and learned how to live it, even when that life led to death. They journeyed, not knowing how the story ended. They leaned in and were willing to add to their understanding of God and what he could do. Instead of saying to themselves, "Our God can't do that," they said, "Why wouldn't our God do something like that?"

A lot of the time, we shortchange the disciples, or at least I do. I think, "Come on, you've been following this guy for six chapters now and you don't even know for sure that he is the Son of God?" But what I fail to remember is that they, too, have been asked to journey with Jesus and that they had to choose to "show up." As compelling as it must have been to see and be with Jesus, I don't think there was some invisible force pushing the apostles to follow him—otherwise, where was everybody else? No, it was a choice for them, just as it was for the Pharisees. I have to give the disciples credit. They were willing to let God open them up to new understandings of faith. They were willing to explore.

And sometimes I'm not. Sometimes I am just busy completing the duties I've been given, or I have other goals for the day. I think, "I don't have time to explore. I'm on this path here, and it's demanding enough." Change and expansion take time and energy, and some days, I'd rather coast around on what I think is the "sure thing."

But, if we're honest with ourselves, we realize that the Pharisees' choice (and the apostles' choice) is our choice, too. We, like them, have to decide on our mode of "follow-ship" when it comes to our faith. We have to decide whether we want to open ourselves up to a growing, changing view of God.

We can choose not to. No feathers are ruffled, and we don't have to figure out what to do with all sorts of new information about God. There are no gray areas here, no paradoxes. Things will be the same as always. The only downside is that we don't yet know what the cost of this strategy will be for our lives. We know what the cost was for the Pharisees; we don't know what it will cost us to "pass" on faith opportunities.

The other strategy for following Christ involves letting God's love lead us to new places and to new understandings of life. We observe and notice life on God's terms just like the apostles did. We follow up those observations not with judgment but with questions. We spend more time with Jesus in order to see what his priorities are. There is cost here, too. There are things about us and about our old ways of viewing the world that will have to go. Gone also are the easy answers and the religious boxes we check off.

But it is worth it! In Romans 8:12-15, Paul, a Pharisee-turned-Christ-explorer, talks about this transition:

> So don't you see that we don't owe this old do-it-yourself life one red cent. There's nothing in it for us, nothing at all. The best thing to do is give it a decent burial and get on with your new life. *God's Spirit beckons.* There are things to do and places to go! This resurrection life you received from God is not a timid, grave-tending life. It's adventurously expectant, greeting God with a childlike "What's next, Papa?"[1]

Who wouldn't want to live listening to God's beckoning Spirit? How exciting to journey with Jesus this way—learning, growing, and seeing life as it really is!

Of course, there is a third option in this choice. It's probably the one I choose the most, and maybe you do as well. It's like my fortune cookie. I say I want adventure, but when an opportunity presents itself, I laugh. I'm too tired, it's too hard, and I just want to reboot. I don't follow through with my cognitive choice. In my head, I choose openness and living by faith; in my day-to-day actions, I often scoot by on my nice and neat pharisaic view of God.

What I've realized is that I'm good at doing spiritual things and at the same time forgetting that life in Christ *is* a spiritual thing. It *is* a dashing and bold adventure, with constant growth, exciting changes, and new chapters not yet written. We need constant spiritual input. We need to quiet ourselves and listen to the Spirit that is beckoning us along unknown terrain. If we're not listening, then we're probably just busying ourselves with completing our religious obligations. We're washing our hands and tithing our herbs just like the Pharisees were. We must not forget how big our God is, that we can never truly, fully know Him, and that we'll get to explore the mysteries of his love our whole lives and beyond.

If we choose the spiritual expedition but find ourselves living like religious hermits, what do we do? How can we keep ourselves open and exploring? Ask yourself these questions every now and then:

- What do I know about God that is different from how I thought about him last year? Last month? Last week?
- Am I comfortable with my view of God changing? Why or why not?
- What has God taught me recently about the world and the events in it?
- What has God taught me recently about my life and the events in it?
- Am I reading the Bible and praying to reinforce my current ideas and experiences or to gain new ones? Am I reading to learn?
- Am I reading the Bible to elicit change?
- Do I expect there to be any effect on my life from the times I spend worshiping God?

Praise be to God that he wants our understanding of him to grow and to be healthy. If we are to look to him with an expectant, "What's next, Papa?" like it says in Romans 8:15, then we must remember that healthy children grow and change. If they don't, something is wrong.

May we commit our lives to the eager exploration of each facet of the character and action of our God. May the Spirit that beckons us lead us on the dashing and bold adventure of life in him.

Discussion Questions

1. Have you opened yourself up to a growing, changing view of God? Why or why not?

2. Prayerfully look back over the list of questions mentioned above. Listen to what God might say to you.

Note

1. Italics mine.

Chapter 11

Perpetual Departure

To see, open your eyes.
To hear, open your ears.
To learn, open your mind.
—quote from a poster displayed in an
English classroom for refugees

Late one August a few years ago, I boarded a plane in Addis Ababa, Ethiopia. It was a routine flight, leaving at midnight and headed for Amsterdam. I wasn't traveling alone. With me were my husband, Matt, and our nineteen-month-old son, Simon Mikias (mih-KEE-us), who had known us for precisely six days.

Our first meeting the previous Saturday morning was memorable but simple. As we climbed the stairs inside the orphanage to the baby room, the director said, "I love seeing if the parents can recognize which child is theirs." *Are you kidding?* I thought. What few pictures we had of Mikias had found their way into my being.

We walked into the room and sat down by him on the floor, just as if we'd done so every day of our lives. I didn't pick him up at first, but just sat beside him. I took him in, and he took me in. I was smiling. He was staring.

He had never left the orphanage once he'd arrived there, at one month of age, except to visit the doctor. He'd seen people with skin and hair like mine probably only a handful of times. Those people had sung to him, rocked him, and left not long after.

And now, here we were, these strange-looking, different-smelling people who seemed friendly enough. We took him out of the orphanage, out the front gate into the world he'd experienced almost exclusively through the baby home windows. We were all experiencing Ethiopia for the first time, it seemed. The three of us went to crowded restaurants where he sat in high chairs for the first time. We went on rough bus rides without a car seat for him. We completed our appointment at the US Embassy and then had quiet afternoons and evenings to ourselves at our guesthouse. Mikias took it all in, wide-eyed and silent. Even now at midnight, when our flight was leaving, he was quietly all eyes.

After a nine-hour flight, we had a layover in Amsterdam. I couldn't imagine what this child must've thought. All of a sudden, his world was like a photograph negative. Skin that was once brown was white. Hair that was dark and curly was blond and straight. Eyes that were brown were blue. Until a few days ago, he understood words people were saying. Until now, he looked like everyone he saw. Now he was completely immersed in a new world, where people only put one layer of clothing on him and tried to feed him unidentifiable bites. If I noticed the change as I stepped off the plane from Addis in Amsterdam, how much more did Mikias?

The next few months, we watched as Mikias amazingly adapted to his new environment. This child seemed resilient. Strap me tightly into a seat and then drive me 65 miles an hour? No problem. Have me sleep in a new room by myself when I'm used to the comforting sights and sounds of others? No problem. Let me get licked daily by a large, fuzzy animal that lives at my new house? I'm rolling with it. Feed me raw fruits, veggies, and cold things like ice cream? Problem. (Mikias hated cold things for the first few months. He'd never really had them before.) He and his new sister Lucy had sibling struggles just like every family that adds a child, and we had our share of adoptive family issues also. But as we watched that first year go by, we knew this transition was complicated. We were filled with simultaneous joy and sadness, for we knew that as we rejoiced, Mikias quietly mourned his old view of the world and the people in it. Each day, though, he'd wake up for more, braver week by week in his new environment.

Mikias is who I want to be: a person who wakes up each day ready to take in the "new," ready to take in fresh information and new scenarios. He is ready for the "more" that this world wants to show him. He doesn't worry that he doesn't quite have a way of classifying the novel information; he just figures it out as time goes on. His whole world is different, and he is diving in headfirst. He is thriving in the midst of the changing equilibrium, and I aspire to be like that.

We see in the Old Testament that it is a choice to accept new worldviews, to leave "home" literally or figuratively, in expectation of God's loving purposes. We see that in the story of Abraham, who left not knowing his destination. We see his daughter-in-law, Rebekah, continue that faithful response, accepting a new nose ring and a new life with people she'd never met. We see Joseph thrive in Egypt as he witnesses God's hand at work to save his family (and the Hebrew people) from famine. We see the Israelites choose to leave Egypt, picking the uncertainties of the desert over the certainties of slavery. And, though they threatened to return to the horrible certainties of Egypt, they never did.

When we read the Bible, we see that *following God is perpetual departure*. We see that *faithful discipleship is letting Christ lead us in consistent worldview change*. Author Philip Sheldrake discusses his idea of perpetual departure within the mystical Christian tradition, writing, "The Christ whose permanent dwelling is in the heart of the human soul is also the one who always travels ever onwards in pilgrimage. The disciple is not left in comfortable and comforting *stasis*, but is thus drawn after Christ."[1]

We see a posture of perpetual departure throughout Scripture and in the life and teachings of Jesus. "An important feature of Jesus' practice," says Sheldrake, "was to push people, not the least those closest to him, away from familiar places into locations they found disturbing."[2] Jesus talked to women—even to women others *didn't* talk to in order to preserve their own reputations. He purposefully traveled outside the traditional Jewish land to spend time with Gentiles. He held children, touched lepers, and took paths through graveyards where the crazy were confined. He did all of this with people following him, taking them to the "uncomfortable places" and helping them understand who the true Center of the world really is.

Many times, people who follow Jesus envy the disciples. We wish we could have been there to see what his followers saw. We think that would make it easier for us to believe and follow. But if we read the Bible correctly, the experience of his disciples is disturbing. Everything was changing, and none of them knew what they were getting into. There were unanswered questions, multiple opinions, and all sorts of new information they had no way of classifying. It must have felt like Jesus had them put their foreheads on a bat, spin around ten times, and then face toward the world as he said, "Run!"

In this way, I don't envy Jesus' followers. I know how uncourageous I really am and how much I like stability. It makes me admire them all the more, however, because they held on through all the "new." They ran, even though they didn't have their balance, knowing that Jesus was leading them. If I had lived in those days, I pray I would have been in the group of disciples who stayed, but, honestly, I could easily have been in the group that said, "This teaching is too hard," and went home to the comfort of their day planners.

This is because, as much as I admire Abraham, Rebekah, Jesus, and Mikias, I don't necessarily like the perpetual onslaught of new and different things. I don't automatically seek a change in equilibrium; I like to keep my balance. I like scenarios that are familiar, meetings with agendas, and food I recognize. I like knowing what I'm getting into and being able to know, if vaguely, how and at what time it will end. Bottom line: I like what I can know. I like what I can control and be sure of because it puts me at the center of everything. It's safe, it's convenient, and I don't have the potential of failure.

Unfortunately, "me at the center" is blasphemous. It is feeding the lie that the world should be how I like it. What I'm learning is that the reality of God's world feels unfamiliar, much like Mikias's new environment felt to him. This journey shouldn't feel stable, because if it does, we might be charting our path instead of letting God do it. Letting God's right-making work set our course puts us in the middle of "frailties and uncertainties." As author Andy Jackson puts it, "Wherever God's people find themselves, there will be worldview issues to confront."[3]

We should pray for open-minded courage as we encounter these worldview issues. We who follow Christ should be found perpetually departing from our own "security" and joyfully engaging the new people, places, and experiences he brings our way. Even though we don't naturally do it, we must learn to receive these new ideas and views of the world. God will stretch us, steady us, and open us up to *his* reality, as Philip Sheldrake describes:

> The presence of God . . . is always strange and elusive, overflowing boundaries into what is "other." This excess, overflow and transgression confronts the human tendency to self-enclosure and to individual self-reference as the measure of everything and everyone else. There is a way of placing "the other" on the margins in reference to the individual person as centre. God is the disruptive action within each person that decentres this illusory centre. . . . God is a radical interruption at the heart of all individual lives that challenges self-containment.[4]

Following Christ, as this author so wisely puts it, is a perpetual departure from myself as the end-all, be-all. It is a willingness to entertain ideas, people, and cultures, not just in a sighing, rolling-your-eyes-for-the-sake-of-the-gospel acceptance, but in a genuine love-embrace of who God has made them and how they can lead you more toward him and his rightness. The difference, again, is in perpetual departure from self. If I am the center, then who *I* am, what *I* know, and what *I* want is what I impart to others. If God is the center, then other ideas, people, and places I encounter have something to teach *me* about *him*.

So we see that *following Christ in perpetual departure means being stretched*. I don't know what this means for you. I *do* know that there is a relationship between the place where you live (or the places you have lived), your values and assumptions, and your perception of reality. I don't know what it will mean for you to begin to get outside of that. But, like me, you need to practice perpetual departure like any other Christian discipline in your routine. This might mean you commit to taking trips abroad regularly, or it might mean engaging in an activity in your own city that stretches you and your thoughts.

When I was younger, I had absolutely no desire to do this. I didn't care *ever* to leave the United States, and I probably even had a small chip on my shoulder whenever I encountered people who had traveled extensively, for whatever reason. It was like they were in an exclusive club, like there were things about the world that they now knew and I could never comprehend. I always felt like saying, "Just tell me what you now know. I'm a fairly intelligent person. I can probably get it, and then I can skip the trip."

But what I didn't understand then is that traveling isn't about knowing trivial fun facts about another place in the world. It's about knowing *yourself* and, still better, knowing *God* in new ways because you were in a completely different environment. No one can tell you what it's like to travel abroad because they can only try to explain—and sometimes they won't or can't—what it was about themselves that they realized while they were gone. Sometimes travelers can't even enunciate what they learned about God on their journey because they can't classify it or fully understand it yet. Once I began traveling, I realized that no matter where you go, departure is about bumping up against your own worldviews, about going to a place where your own presuppositions (ones you weren't even aware you had) jump out at you around every corner.

This departure, this stretching, isn't really about distance. It's about giving God the opportunity to grow your understanding of the world he made and how he wants to set it right. It's about letting him change your view of your own city and see where your *own* culture needs to be set right. It's about realizing that God often works in the journey, in the space between "fixed places." Departure means letting God be the center and following Christ out of your self-containment and into the broad, spacious truth of life. As Sheldrake says, "discipleship simultaneously demands a place and an 'elsewhere', 'further', 'more'."[5]

This kind of life sounds great, doesn't it? Especially if we use words like "fresh adventures" or "new insights." We want those things for our lives, or at least I do. It's hard for me to follow when perpetual departure gets messy. I like nice and neat; I don't like complex. I like planned success; I don't like the possibility of failure or embarrassment. But what we see in Scripture is that *following Christ in perpetual departure not only stretches us but also involves risk.* Sometimes it involves being okay with being uncomfortable. Sometimes it involves a willingness to look silly. But almost always, perpetual departure is not as heroic as it sounds from the outside.

To follow Christ means to depart perpetually, even though we don't have rights, don't have an easy road ahead, and don't fully understand everything. It means we depart from our limited worldview and trust that "the earth is the LORD'S and everything in it" (Ps 24:1, NIV). To depart perpetually means that we discipline ourselves simply to "go," to exit ourselves and trust in God when we encounter what we can't understand. As theologian Simone Weil said in her book *Waiting for God*, "It is necessary to uproot oneself. Cut down the tree and make it a cross and carry it forever after."[6]

Mikias's bed is right by a window on the front of the house that overlooks our street. Early in the mornings when I pass his room, I sometimes find him gazing out the window, his chin propped on one fist and the thumb from the other hand in his mouth. I would give millions to know what is going on behind those dark eyes. Some mornings I'll go in and ask him if he wants to get up with me and he'll say "Not yet," and turn back to the window. Other mornings I'll wake up to his voice announcing the sighting of his favorite vehicle: "Mom! I see an orange school bus!"

When he looks out his window here, he sees paved streets, mailboxes, and houses with driveways and basketball hoops. When he looked out his window in the orphanage, he saw vines and grasses, piles of rocks, and buildings made of cinderblock with sheet-metal fences. He saw goats, donkey carts, trucks, and pedestrians sharing the unpaved mud "roads." The only commonality would seem to be the sky; everything under it has changed.

Yet, when I see him taking it all in, lying window-side in his Superman pajamas, I see who God wants me to be. God wants me to be a sojourner who is willing to learn, accept, and even enjoy the uncertainties. I see Mikias and I believe he *is* a superhero. He teaches me that it is only when we follow Christ in perpetual departure that God can connect us to himself and to each other, our forever family.

Discussion Questions

1. How might God want to stretch you? What is the "elsewhere," "further," or "more" to which he might want to take you?

2. What are some of the "frailties and uncertainties" you are facing in your current place?

Notes

1. Philip Sheldrake, *Spaces for the Sacred* (Baltimore MD: Johns Hopkins University Press, 2001) 141.

2. Ibid., 69.

3. Andy Jackson, "To See With New Eyes," in *Pathlight: Toward Global Awareness*, ed. Meg Crossman (Seattle: YWAM Publishing, 2008) 65.

4. Sheldrake, *Spaces for the Sacred*, 67–68.

5. Ibid., 31.

6. Simone Weil, *Waiting for God* (New York: Harper & Row, 1973) 7.

Chapter 12

Transcending Particularity

The day you can pack your church on my camel is the day I might consider Christianity.
—quote from a nomadic Somali man[1]

The building that houses my church is probably not going to be a church much longer. The building has a huge "FOR SALE" sign on it right now. The congregation has plans to relocate and build on a new piece of land in another part of our city. It is a plan long overdue, actually, because of lack of classroom and parking space. Most church members park blocks away each Sunday and either ride a shuttle or walk to the services.

There have been parking problems ever since our church began, although back in 1839 when the church was founded, they were the horse-drawn-wagon kind of problems. For a while, there was no building; the group of fifteen people met in homes in the winter and in the open air in the summertime. The first building was made of logs, used as a hospital during the Civil War, and eventually burned in the conflicts that occurred there. Another building was built near the present-day cemetery at a cost of $1,500, and the congregation worshiped there until the railroad came to town in 1875.

Then it was decision time: the entire town of Blue Springs decided to move a couple miles over in order to be close to the railroad line, and the church had to decide what it wanted to do. Move or not move? Embrace tradition or progress? Eventually, in 1879, the church decided in favor of the move. The entire church structure was rolled on logs pulled by a horse team to its present location on 15th and Main Street. New hitching posts were built, and the congregation moved forward with the town. In 1922, a brick structure was constructed on the same site, and we use this building today for worship each Sunday, along with the town's old library building (which is the student building across the parking lot), and a couple of other additions that came in 1956 and 1977.[2]

The "FOR SALE" sign is in front of the steps of the original building. I have to say I am a little hesitant to see what will happen to the building itself once it is sold. That one corner of the universe holds an inestimable number of memories. The brick buildings and the land they sit on are virtual filing cabinets of stories. There are active drawers and dusty inactive ones, storing the life and faith of over a century's worth of people who lived and died in this part of the world.

I would be sick if one day I drove down the street and found a Target store on that corner. It's obviously a good thing that I am not on the church committee in charge of the property sale. I fully agree with the move: the current buildings are inconvenient to those with disabilities, have too few bathrooms, and have been renovated time and again. A change is needed; it's just hard.

I could go on from here lamenting progress and commercialism. It would be easy for me to let nostalgia write this chapter. Change—even change that I believe is necessary—is so complex. Why is that?

It might be because I attach things to my memories. I've always put tangible things around me that remind me of intangible ideas or stories. It never occurred to me that some people *don't* do this until I read this on a personality test I took: "Sees objects as mementos." I have kids' handprints here, a sticky note with a quote on it there, and funky pieces of ceramic fruit in my kitchen that were in my Grandma's kitchen. I have probably fifty yards of polyester fabric from the 1960s from Matt's grandmother that I use for random projects. No wonder we have to make cleaning out the basement storage area a biannual occurrence around the Sprink house: I would hang on to objects until memories literally crowded the four of us out.

If we're honest with ourselves, we all do this somewhat, don't we? We attach objects to inanimate ideas. In some ways, we're trained to do it by our culture of commercials: think car and shoe ads. In other ways, this connection to objects comes naturally as we grow and assign meaning to them in order to make sense of our world.

I was recently talking to a friend who was preparing for a garage sale. She told me that as she and her daughters, Avery and Kate, were going through their toys, it was painful for them to get rid of any of their Barbie dolls because they all had names. Finally, after a few thoughtful moments, the two girls picked out two Barbie dolls that could be sold on one condition: they must be sold to the same person because they were sisters. We can smile at this because it's not happening to us, but on some level, we *all* attach objects to our ideas. We *have* to, especially when we're talking about big, inexpressible ideas like *love* or *family*. Just like my little friends Avery and Kate had attached *sisters* to their Barbie dolls, I grew up in the brick building on 15th and Main with certain hallways, light fixtures, and stained-glassed windows that my brain labeled *faith*.

Our faith, of course, will take on particulars like people, places, and routines. Songs, images, verses, and faces. These are cultural forms, objects to which we attach meaning in order to express the inexpressible. We all do this, and it's okay. We *have* to. How else could we begin to understand the story of God and his interaction with us?

God himself knows it is difficult for us to comprehend all this. Why else would he have given us objects-as-mementos like the ark of the covenant or the temple? Why else would he *himself* enter the particulars of our world to become human, if not so that we could actually experience him? Why else would he have told parables or given us the bread and the wine? Particulars like objects and rituals that represent our faith are not evil, just like the mementos I have around my house aren't evil. They're symbols that provide meaning and connection in life.

The problem is when we get it backwards, when we let the cultural forms *determine* the scope of our faith. Or, put a second way, the problem comes when I fill my house so full of mementos that we no longer have room to live in it. Or, put a third way, it would be wrong of me, and wrong of my whole church, to let nostalgia get in the way of the actual life of the church, to let it hinder the new things that God wants to do in this particular congregation.

We *are* to prize tradition. We *are* to focus on the particular facets of our places, but it is a precarious balance. We can't focus so much on our own memories or preferences that we as a church decide it's fine to have a parking lot where people drive through on Sunday and leave because there are no more spaces. At that point, we'd be letting our memories crowd out space for actual people. At that point, our definition of faith would look like "Jesus + my own faith experiences" or "Jesus + 15th and Main."

The Bible, especially the New Testament, frees us from any kind of "Jesus + _____" faith. If we look at the life of Paul, we see someone who spent his life fiercely opposing anyone who would add cultural requirements to faith in Christ. Of course, we know the beginning of this story: Paul was fully embedded in the first-century Jewish culture and a Pharisee to boot. His life revolved around his Jewish culture and its regulations. He encounters

Christ on a road to Damascus, where he is traveling to squelch the new Jewish sect called "the Way." We read of this in Acts and wonder, what now? The railroad has come to town, and Paul has to decide whether to move or to stay put. What becomes of Paul and the particulars of his previous faith experience? What happens when change comes to all the Jewish religious forms he is now defending? Do the religious particulars of Paul's life, the way he has understood and related to God before now, fly out the window?

Biblical scholar Ben Witherington talks about this in his book *The Paul Quest*. Paul did not renounce his culture, Witherington says, but he let his interaction with Christ inform his culture after his conversion. His relationship to Christ put his cultural identity, or the objects and rituals his brain had labeled *faith*, in proper perspective: "Paul's Jewish heritage cannot simply be ignored or dismissed as a key to understanding his identity."[3]

After careful analysis in his book, Witherington says Paul's resocialization on the Damascus Road consists of the following (according to Paul's own descriptions): (1) Paul happily claimed his Jewish heritage post-conversion. (2) He did not renounce this Jewish faith as something wicked when he became a Christian. (3) He was unashamed of his background and record as a Pharisaic Jew, with the exception of his persecuting Christians (see 1 Cor 15:9; Gal 1:13, 23; and 1 Tim 1:13). (4) It is obvious that something has surpassed the importance and pervasiveness of his former worldview. "The heritage of Israel, however, is by no means being renounced," Witherington writes. "In Paul's view, it is being claimed and fulfilled by Jews and Gentiles united in Christ. The promises of God to Abraham are fulfilled in Christ."[4]

When we look at Scripture, we see a Paul who maintains and is proud of his Jewish background (the particularities of his faith experience) and at the same time now sees the centrality of Christ as paramount. Is it okay for Paul to be Paul, to have his own religious and cultural particularities in his faith journey? Yes.

So what's the problem? The problem is not so much a "what" as a "who." We see in Acts 15, and in our own lives, that it is hard for church people to remember that "Jesus" alone is central to a life of faith, not "Jesus + my religious experiences."

Some religious men from Judea came to Antioch, where Paul and Barnabas were seeing God transform the lives of *all kinds of people*, and they began teaching that in order to be saved, the formula was Jesus + circumcision (one of the requirements of the Law of Moses). Salvation through Jesus was only for the culturally Jewish, in other words. Paul and Barnabas could not accept this simply because they had seen Jesus act in the lives of both Jews and non-Jews. So they went to Jerusalem to consult with the apostles and elders on the matter.

When Paul and Barnabas stood before them and told of the ways that God was breaking in to all the different people they had encountered, we think, this sounds familiar. And Peter thought the same thing. For it was only a few chapters earlier in Acts 10 that God led him to the house of Cornelius, a Gentile (a visit that was against the Jewish law).

While there, Peter witnessed the outpouring of the Holy Spirit on this believing household and said to them,

> It's God's own truth, nothing could be plainer: God plays no favorites! It makes no difference who you are or where you're from—if you want God and are ready to do as he says, the door is open. The Message he sent to the children of Israel—that through Jesus Christ everything is being put together again—well, he's doing it everywhere, among everyone.[5]

And what happened to Peter? He was criticized by the religious people in Jerusalem for breaking a cultural law. The Jewish Christ-followers had forgotten that faith was "Jesus" and not "Jesus + Jewish law." So Peter explained, "God came to them when they heard the story of Jesus. Who am I to control what God is doing?" As the church people listened to Peter, they let this new information simmer in the crock pot. "And then, as it sank in, they started praising God," writes Luke in Acts 11:18. "'It's really happened! God has broken through to the other nations, opened them up to Life!'"

How amazing this realization must have been! It was so revolutionary that Peter, when he was present at the Jerusalem council and listening to Paul and Barnabas's stories, remembered. We read in Acts 15 that he takes the floor in support: he tells his story of Cornelius. He reminds fellow church members that the story is for all people, and he challenges them: "Why are you now trying to out-god God?" (Acts 15:10).

James and the church eventually come to the understanding that their Israel-centered idea of Jesus' message is just one piece of the puzzle. Their Jewish cultural forms are acceptable as they express their devotion to Jesus, but those forms are just *one* part of a much bigger picture. Peter, Paul, and Barnabas say that God is including so many others in his restoration work as he completes the whole puzzle—the picture of the kingdom of God.

God was opening up the offer of real life to everyone, and the early church, with its ethnocentrism, had inadvertently narrowed the focus of the kingdom of God. Paul fought against this ethnocentrism for the rest of his life; to Paul, the good news of the work of Christ was that it is effective for *all people*. Through his ministry of presence and his ministry of correspondence, he pressed toward the goal of uniting the Jerusalem church and the Gentile churches *not* under cultural identity but under the life-restoring work of Christ.

We can't possibly know how difficult it was for Jewish Christians to try to see how non-Jewish people could know God in other ways. After all, it took a vision of Christ for Paul to grasp it and a vision from a rooftop for someone like Peter to understand. To think that others could connect with Jesus through cultural forms other than theirs shattered their worldview and threatened everything they knew and loved about their faith. What is more, the way they limited the gospel's scope might not have occurred to the followers of Christ had it not been for God working through Paul. In an interview with Peter Jennings, author Karen Armstrong said, "Were it not for Paul, and his insistence to include non-Jews, those who followed Christ would probably have remained a small messianic sect within Judaism."[6]

But actually, when I think of my home church and its changes in location over the years, maybe we can identify with the Jewish Christians a little more than we first thought. Change is hard. It's hard to understand the ways that others relate to God. It's a lot of work to give space and respect to the faith cues of others. It's difficult to see the importance of their "objects as mementos" when we are so used to our own.

When my interactions with God are all I know, and when I'm only around others who interact with him in the same ways, it is easy to begin to attach those faith cues to Jesus. And as we discussed before, those things are not bad; they serve a good purpose. We get in trouble today, just as the early church did, when we begin to narrow our understanding of what God is doing in the world based on our own experiences. When we begin to say to others, "Your faith in God must look like mine, or it's not *really* faith," we have turned into the pharisaical first-century Christians who began to define faith as "Jesus + my religious preferences."

If you don't think this happens today, consider the quote at the beginning of this chapter from a nomadic Somali man to a Christian missionary (a quote we heard from a Somali pastor). Why else would an intelligent man think he was unable to experience the Christian God unless he came to a certain place to do so? In this way, the well-intended missionary *inadvertently* excluded this Somali man from Jesus' message because of his nomadic lifestyle.

And that's really the point. The point is that we Christ-followers inadvertently attach our faith experiences and preferences to the work of Christ, which often excludes or confounds those who may want to experience Christ for themselves. For example, I would never say that it matters where my church meets; of course, people can worship God anywhere, even not in a building at all, like the founders of my church did. But if I let my affection for our building keep me from embracing (with my actions) the new ways that God is moving in our community, I inadvertently limit the scope of God's message in Blue Springs, Missouri. What is the cost of my doing that?

This was the Apostle Paul's point. There is to be no dependence on cultural identity or cultural preferences in the Christian life. There is no "in group." God's love, according to John 3:16, is not only for those who are like us. It is for *all*. Of course, of course, we say. God loves everyone. But what is implicit in this claim that we do not recognize? What characteristics do we assume about a person whom God loves? A Western worldview? A Judeo-Christian work ethic? Family values? Social success?

We must learn from the tendency of humanity to ground one's faith in particularity instead of in Christ. When we read Acts, we as Christ-followers must examine the nature of the work of Christ and celebrate our freedom from culture-bound forms. We need to examine constantly what is central to the work of Christ and discern what parts of our faith are cultural or faith-based forms.

In recognizing these distinctions, in looking into what is truly central to the message of Christ, we can come to a fuller understanding of what God is truly doing to restore the world. As professing Christ-followers, we must constantly ask ourselves, will we trust in the universality of our message? Are we willing to trust *only* in the work of Christ and its effect on our lives?

Do we believe in, and really *lean into*, the universality of our message? Will we let the work of Christ stand on its own, or will we require others to conform to who we are? Or are we, like Paul, willing to risk and willing to trust in the universal nature of Christ's work, helping others to creatively discover the effect of Christ's work on their own lives and on their own culture? May we as the church be found faithful to the message that was wide enough to include us.

Discussion Questions

1. What are some of your faith cues?

2. Do you believe in and lean in to the fact that Christ is for everyone? Do you recognize Christ's work in the lives of others, even when it looks drastically different from the way he works in your own?

3. How do you help others discover Christ's creative work in their lives?

Notes

1. The "God can't fit on my camel" story was shared with us by some Somali friends.

2. The history of our church and of Blue Springs is compiled from information collected from church members, the First Baptist Church "1994 State of the Church Address," and http://www.rootsweb.ancestry.com/~mobshs/baptistchurch.html (accessed 17 August 2011).

3. Paul still figures time according to Jewish festivals as stated in 1 Corinthians 16:8 and gives his "credentials" to those asking in 2 Corinthians 11:22 and Philippians 3:4-6. It also seems to be significant that he says in the same 2 Corinthians passage about his heritage, "so am I" instead of "so was I" (Ben Witherington III, *The Paul Quest* [Downers Grove IL: InterVarsity, 1998] 55).

4. Witherington, *Paul Quest*, 55.

5. Acts 10:34-36.

6. Peter Jennings, *Jesus and Paul*, ABC Interview with Karen Armstrong, aired 5 April 2004.

Chapter 13

Our Communal Place

We fail in our duty to be human when we forget that we are everyman and everywoman.[1]

We walked into what looked like an airport terminal. There were rows and rows of attached chairs, filled with waiting men, women, and children. There were people standing, sleeping, and arguing with other exasperated people working behind counters. We saw long corridors with numbers and arrows pointing this way and that. People were being carted here and there on golf carts, and the drivers had a hard time making their way around groups of people who looked like they had camped on the floor for the last week. All sorts of languages were being spoken, and everyone seemed irritated and exhausted. If it weren't for the lack of airplanes, I'd have thought I was at DFW or JFK.

I wasn't. I was in a hospital in Johannesburg, South Africa. It was my first day there, and a friend was taking my husband and me on a tour. Eventually, we would spend weekly time there as members of a support group for people living with HIV and AIDS, but this was our first visit, an orientation of sorts.

I was introduced to a nun, Sister Mary, who served as a volunteer counselor. She told me to come with her, and I did not question but followed her alone. She ushered me into a small exam room to the side of the terminal-like waiting area and told me to wait there. I was happy to have a minute alone to process. I hadn't been in the country for more than a couple of weeks or so, and I was still taking everything in. I looked around the exam room, a fairly familiar-looking place with health posters on the wall, swabs, and sterile metal furnishings. I sat in one of three chairs in the dimly lit room.

In the next minute, Sister Mary entered and sat in the middle chair, offering the one across from me to a young African woman named Nancy. They spoke English, which was a grace for me, and Sister Mary introduced the pretty young girl. Her hair was covered with a beret, and she looked like she was in her twenties, as I was at the time. I remember noticing her beautiful, dark eyes.

After a few exchanges of a general nature, Sister Mary kindly but candidly told Nancy that her blood work had come back and the lab work showed her to be HIV positive. As I listened, I learned that Nancy and her spouse were expecting and she had come in for a regular prenatal checkup. Sister Mary began talking about cell counts, using formula instead of nursing, and why it was important to use condoms even after infection. I heard less than half of what she was saying; I can't imagine Nancy heard anything. She asked no questions but simply sat, shedding quiet tears. Sister Mary left the room briefly. I don't remember why she stepped away, only that it felt like a soundless lifetime. Nancy spoke English, so it wasn't for a lack of language that we were silent. It was because there were no words.

I could only stretch my hand out and place it on her knee. We sat like that until Sister Mary returned. She worked out a plan for follow-up appointments with Nancy and encouraged her to eat a nutritious diet as she left. The visit was over quickly, as more people were waiting. There were no explanations in this part of the hospital, no time to ask "How?" This seemed abrupt to me, and at the same time like grace. There was only loving, practical action to help the hundreds of Nancys who waited in row after row of chairs. I later learned that the doctors, nurses, and everyone else working in the clinic were volunteers.

That day, in that place, I sat across from myself. I've been trying to process this brief interaction for the past ten years. Even now, I fumble it when I try to explain it to people, and it sometimes mistakenly comes out like I'm saying, "If I had been born in Tibet then I probably would have been a Buddhist." That wasn't it. There in the hospital, I didn't feel lucky that I had been born in America; I felt horrible that *anyone* had heard the terrible news Nancy heard. Sitting in that place, I was sitting across from South African Holly, and *that* Holly wondered why the educated white American who reached across the room and put her hand on my knee hadn't truly understood, until now, what her own generation was going through. American Holly's offense was not her silence that day; it was her prior indifference.

In the journal I used while in South Africa, I scribbled this Dietrich Bonhoeffer quote:

In view of such terrible human catastrophes the arrogant spectator attitude of a judge or know-it-all is no longer valid for the Christian. Rather what really counts here is that we realize this one thing: these events took place in my world, the world I live in, the world in which I commit sin, in which I sow hatred and unkindness day by day. These events are the fruit of what I and my family have sown.[2]

There, at the weekly support groups we began helping with, no one spent time on infection stories. In fact, many people who came and went assumed we were HIV positive as well. It didn't matter to us. What was really important, according to Bonhoeffer's account, was recognizing our common humanity and choosing to show up. Theologian Simone Weil puts it this way:

The love of our neighbor in all its fullness simply means being able to say to him: "What are you going through?" It is a recognition that the sufferer exists, not only as a unit in a collection, or a specimen from the social category labeled "unfortunate," but as a man, exactly like us, who was one day stamped with a special mark by affliction.[3]

Our place before God is first and foremost a communal one. Our individual cultures and histories are important, to be sure. However, they must not detract from our common humanity. They must enhance it instead. In Christ, we have true spiritual freedom. As Americans, we have political freedoms. But what is the purpose of these freedoms? One author writes, "Though we receive the freedom of the children of God, surely that freedom consists of more than accumulating the rights and privileges by which we may fulfill ourselves."[4]

The apostle Paul wrote about the correct use of freedoms in his first letter to the Corinthian church.

Though I am free and belong to no man, I make myself a slave to everyone, to win as many as possible. To the Jews I became like a Jew, to win the Jews. To those under the law I became like one under the law (though I myself am not under the law), so as to win those under the law. To those not having the law I became like one not having the law (though I am not free from God's law but am under Christ's law), so as to win those not having the law. To the weak I became weak, to win the weak. I have become all things to all men so that by all possible means I might save some. I do all this for the sake of the gospel, that I may share in its blessings. (1 Cor 9:19-23, NIV)

Should we find ourselves with unimaginable religious and political freedoms, writes Paul (and as Christians and Americans, such is the case), *it doesn't stop there*. Should we ever find ourselves sitting in a room across from a young South African woman, the message of Christ necessitates connection. It demands self-sacrifice as Christ gave on the cross. One scholar, Edgar Krentz, writes, "The boundary to personal liberty is the death of Christ on behalf of the weakest, poorest, most marginalized person."[5]

What keeps us from a communal mind-set? How is it that I read online statistics about the millions of Nancys out there and then forget to take them into consideration when I make choices about my life? I can think of three possible reasons, and I'm sure you could add more.

First, I think we are consumed with the details of our lives, forgetting that waking up each day is about Christ and the way he works to make things right in the world. Paul writes further about the temptation to forget in Galatians 5:13, NRSV: "Do not use your freedom as an opportunity for self-indulgence, but through love, become slaves to one another." Are we guilty of going about our days, presuming upon the graces we've been given, and using them to make ourselves look good? Scholar Lyle Vander Broek writes,

> The Paul who was called by the risen Christ has been freed from an agenda of self-promotion. He no longer has to carefully guard his own rights or protect his social status. Paradoxically, Paul is free to give up his own freedoms. Christian freedom, in the fuller sense of the word, means the freedom to follow Christ, the freedom to love and serve those around us in a way that would have been impossible in our previously self-centered world.[6]

I love this definition of freedom. It is almost as if Vander Broek is describing our self-centered thoughts and activities as clutter that needs to be cleared so that more space will exist in order to love others. I can almost hear the apostle Paul saying, "How much more time would you have if you didn't fret in front of the mirror, making sure everything matched and every hair was in place? Sure, I'm no beauty queen, but who has time? There's a *whole world* out there!"

So what is the solution? We de-clutter. Only when we can free our calendars and our thoughts of the self-aggrandizing tasks will we be able to make space for others in our lives. God calls this being willing to be a priest to others, as it says in 1 Peter 2:5, "Present yourselves as building stones for the construction of a sanctuary vibrant with life, in which you'll serve as holy priests."

Some who follow Christ call this idea the Priesthood of All Believers. In the tradition in which I was brought up, the idea is this: "All Christians belong to the priesthood. They are priests in their freedom to come directly to God through no other mediator than Jesus Christ. They are also priests to one another, called of God to fulfill priestly ministry amid the brokenness of the world."[7] Yes, making space in our calendars is a sacrifice, but I believe freeing our schedules and our minds of our concerns is a protective practice that keeps us from becoming re-enslaved to ourselves and to others.

In his book, *The Longing for Home*, Frederick Buechner writes that when we have freed ourselves to be priests or saints to others, then we have truly found our own place before God.

> I receive maybe three or four hundred letters a year from strangers who tell me that the books I have spent the better part of my life writing have one way or another saved their lives, in some cases literally. I am deeply embarrassed by such letters. I think, if they only knew that I am a person more often than not just as lost in the woods as they are, just as full of darkness, in just as desperate need. I think, if I only knew how to save my own life. They write to me as if I am a saint, and I wonder how I can make clear to them how wrong they are.
>
> But what I am beginning to discover is that, in spite of all that, there is a sense in which they are also right. In my books and sometimes even in real life, I have it in me at my best to be a saint to other people, and by saint I mean life-giver, someone who is able to bear to others something of the Holy Spirit, whom the creeds describe as the Lord and Giver of Life. Sometimes, by the grace of God, I have it in me to be Christ to other people. And so, of course, have we all—the life-giving, life-saving, and healing power to be saints, to be Christs, maybe at rare moments even to ourselves. I believe that it is when that power is alive in me and through me that I come closest to being truly home.[8]

When we are willing to live communally, to align our lives alongside each other, we find ourselves more at home in the world than we ever thought we could be.

A second, related reason we fail to look at the world compassionately is that we tend to prize individualism above all else. For instance, when I read about unity in John 17 when Christ is praying for the unity of the world, I have no problem with this. It is a wonderful idea. Who would disagree? We like the idea of unity, of connection with others. Sometimes, however, we act like a small child who has just learned to talk, having the newfound ability to say to a helping parent, "I can do it by my*self*!" We think that living as individuals and having the freedom to do things the way we want is the ultimate form of existence. We forget that there are parts of us that need to know others, and that we must give ourselves to them in order to be fully ourselves.

"The impulse of sin," writes Silas Bishop, "is to choose to be self-sufficient, self-focused, and independent of others, seeking satisfaction outside of relationship."[9] How interesting this definition is, composed of words that we in the West are conditioned to embody. Self-sufficient! Self-focused! Independent! Who wouldn't want those words on a résumé? Who wouldn't want children with these qualities? How could they relate to *sin?*

Maybe the problem is that we stop with independence. Maybe the harder calling, maybe our true freedom, comes with interdependence. "To be free," reads a Nelson Mandela quote on the wall of South Africa's Apartheid Museum, "is not merely to cast off one's chains, but to live in a way that respects and enhances the freedom of others."[10] Our true freedom comes in recognizing our common humanity, in being community to others so that they can inform our journey.

How appropriate, then, that when we look to God, we see that community is what God is and what he is about. "God is the social Trinity, the community of love," writes theologian Stanley Grenz.[11] Another author writes that this interdependence "is so clearly at the heart of what it means for God to be the Trinity—what it means for God to be God—that it is not an oversimplification to say, 'God is community.'"[12] If we see this social, interdependent reality in the person of God, how much more should we set as our goal this divine nature, this sharing of life with others! We also know that Christ promised to be in our midst when we gather together, moving us toward an understanding of the communal nature of our faith journey. "The *people* of God are the place of God," says Philip Sheldrake.[13]

What is the solution to our obsession with independence? We must keep before us the idea that *who we really are is who we are when we are in relationship to others.* For example, as I am the mother of two young children, I am frequently interrupted by my kids when having a conversation or a meal with someone I don't know well. Sometimes I later think, "Boy, that person has no idea who I really am because we didn't have much time for a *real* conversation." But if I am thinking communally, I shouldn't long for time to "be the real Holly" who is unhindered by children. After all, the real Holly *is* the one with children in tow, the one taking five trips to the bathroom with them during a meal at a restaurant. Who I really am is how I act when I am in relationship to my family, friends, and the world. Who we really are is how we relate and connect to people everywhere. As Fyodor Dostoevsky wrote in *The Brothers Karamazov*, "A man's true security lies not in his own solitary effort, but in the general wholeness of humanity."[14]

A third reason we don't live with others in mind, I believe, is that we don't think we need each other. It is easy for us, when we stay in one place and have lived at this certain time in history, to believe that what we know is all there is to know. There's no reason to know anything we don't already know, we think, so we begin dismissing things and people. We close ourselves off, sometimes intentionally, because it makes life easier and there are fewer questions or gray areas. We forget that others don't see the world like we do, and so the incredible benefits of learning from others disappear.

But we do need each other. Biology tells us that absolutely nothing lives in isolation. We are already connected far more than we know. These connections, things like the need for food and clothes, are not what Jesus meant when he prayed for our unity. We need each other's stories. We need each other's places to help us understand our own. We need to look outside our own interpretations of life and listen to those of others. Jesus' teaching, says author Walter Wink, holds that "there is something of God in everyone. There is no one, and surely no entire people, in whom the image of God has been utterly extinguished."[15] Shouldn't we who follow Christ desire that others, who have the same Holy Spirit we have, speak into our lives and into our cultural contexts?

Listening to the thread of Nancy's story is the solution. We find that our obedience is not about doing but about being, about walking beside a person and listening. What we hear is what will combat our individualism. When we hold the babies who have no parents, when we sit with the dying, when we hear stories of the way God is working to make things right, we are living sacrificially, as Christ did. Our discipleship—not just that of others—is affected by our choice to be present in a place and to be part of a community. "Radical decisions in obedience are of course the stuff of biblical faith," says Walter Brueggemann, "but now it cannot be radical decisions in a private world without brothers and sisters, without pasts and futures, without turf to be managed and cherished as a partner in the decisions."[16]

May we, the people of Christ, choose to live as he does, in holy communion with God and with others.

Discussion Questions

1. Where or when have you sat across from "yourself"? When in your life have you been reminded of our common humanity?

2. How might God ask you to de-clutter, to clear out space in your life so that you can attend to the needs of others?

3. What do *independence* and *interdependence* look like in your life?

Notes

1. Stuart C. Bate OMI, ed., *Responsibility in a Time of AIDS* (Pietermaritzburg: Cluster Publications, 2003) 42.

2. Dietrich Bonhoeffer, Geffrey B. Kelly, and F. Burton Nelson, *A Testament to Freedom: The Essential Writings of Dietrich Bonhoeffer* (New York: HarperCollins, 1990) 233.

3. Simone Weil, *Waiting for God* (New York: Harper & Row, 1973) 115.

4. Patrick J. Wilson, "The Freedom of Slavery," *Christian Century* (January 1994): 43.

5. Edgar Krentz, "Paul: All Things to All People—Flexible and Welcoming," *Currents in Theology and Mission* 24 (June 1997): 243.

6. Lyle D. Vander Broek, *Breaking Barriers: The Possibilities of Christian Community in a Lonely World* (Grand Rapids MI: Brazos Press, 2002) 95.

7. Bill Leonard, pamphlet, "Priesthood of All Believers," *Foundations of Baptist Heritage*, Southern Baptist Historical Society, 1989.

8. Frederick Buechner, *The Longing for Home* (San Francisco: Harper, 1996) 27–28.

9. Silas Bishop, "Community as the Goal (And Therefore Means) of Mission," *Truett Journal of Church and Mission* 1 (Fall 2003): 9.

10. I read this quote on the wall outside the Apartheid Museum in Johannesburg, South Africa. It can also be found in Nelson Mandela's autobiography, *Long Walk to Freedom*.

12. Quoted in Bishop, "Community as the Goal," 7.

13. Philip Sheldrake, *Spaces for the Sacred* (Baltimore MD: Johns Hopkins University Press, 2001) 37.

14. Fyodor Dostoevsky, *The Brothers Karamazov* (London: Vintage, 1992) 303–304.

15. Walter Wink, *The Powers That Be* (New York: Galilee, 1998) 178.

16. Walter Brueggemann, *The Land: Place as Gift, Promise, and Challenge in Biblical Faith*, 2nd ed. (Minneapolis: Fortress Press, 2002) 199.

Chapter 14

Spacious Relationships

There is no such thing as a physical geography of anywhere divorced from its human geography.[1]

A picture of a man named Mohamed hangs on a wall in our home. It's in a grouping of photos from different places we've been such as South Africa and Morocco, but guests usually ask about Mohamed's photo first. This is probably because Mohamed, a traditionally dressed Muslim man with a thick, black, no-mustache beard, looks like the kind of guy you'd see on the evening news or being body-searched at a TSA checkpoint.

I met Mohamed in September 2001. He walked into my English as a Second Language class one evening, quiet but ready to learn. He was from India, born in Shimla, which is near the India-Pakistan "border." He knew four different languages but wanted to improve his English. A friend had recently brought him to the United States to be the imam, or pastor of sorts, at the local mosque. His family, at that time a wife and a daughter, was back in India. He had a solemn demeanor, but when he talked, his voice was surprisingly loud. He prefaced just about everything he said with a nasal "Ahhh" sound, and when he finished talking, a thin, pleasant smile flickered across his face.

He came every Sunday except for the times when sunset conflicted with class time, as he was leading prayers at the mosque. He became friends with my husband and me over time; I believe the first invitation was from him for dinner at his home, which was a room built onto the side of the mosque (which was built onto the back of a convenience store).

The three of us shared many meals together at each other's houses. Some were incredible curries he made himself that we ate off his small coffee table. Other meals were from my just-married-still-learning-to-cook menu eaten in our little rented house. Our Hindi or Urdu was nil, and his English vocabulary was limited, so we often ate in silence, trying to talk until we reached the point where we just smiled and shrugged because no one knew the words.

We had friendship, but we couldn't make out the whole picture. Each time we were together was an attempt to connect the dots and see more of who each other were.

One memorable meal was a barbecue of sorts that we ate off card tables outside Mohamed's apartment in a fenced-in driveway/courtyard. Matt and I, along with the other ESL teachers, were to eat with Mohamed and several of the men from the mosque. As the food was prepared and dusk fell, we chatted with the men and watched Mohamed flit around doing the host's preparations. I noticed a three-by-six section of the driveway that had not been paved; it was a flower garden with neatly tended roses and other small blossoms.

The inside of his apartment was sparse but neat. On his wall, he had a map of the US and a dazzling electric picture of a waterfall in which the water moved when it was plugged in. He carried one photo of his family with him; it was a photo of the three of them that reminded me of a solemn photo with Santa. He showed it a lot, talking of how much he missed them and how he couldn't wait until he could bring them to America. He frequently said, "It not good for husband to be apart from family."

We shared another memorable barbecue in our much-less-manicured backyard, when Matt graduated from seminary. Friends from church, professors from seminary, and students from ESL, including Mohamed, sat on lawn chairs eating barbecue and watching others take a whack at a piñata we had hung from a pecan tree. I wondered what Mohamed thought of being in a yard full of foreigners who wanted to blindfold him, spin him around, and hand him a bat, but all he said after the party was, "This very good program."

We had interesting talks about what was important to each of us, even with our limited communication. He knew we were in seminary and that Matt worked as a minister, and we knew he led prayers and taught Arabic from the Qur'an to the mosque's children throughout the week. He told us he was led to that profession by a vision he once had when he was deathly ill. He talked about how he liked America and how he felt Americans accepted Muslims. We talked about visiting India together. We talked about the things we had in common from our Scriptures, and we prayed together before our meals.

Years went by. We moved to South Africa and then back. He returned to India for a time and then moved to San Antonio to a different mosque. We continued to share meals—sometimes at the Denny's near his apartment when we were in San Antonio and sometimes when he was visiting in Waco. One time, he called to tell us his son had been born in India, and we celebrated together over the phone. We called to tell him when our daughter was born, and the next time he came into town, he brought a doll for her. During one visit, I casually admired an Indian sari, and Mohamed, surprised, said, "You *like*?" The next time he was in town, he brought me one.

Over the years, as I learned more about Islam through other friends and through study, I became aware of the ways Mohamed put up with us, overlooking our cultural faux pas. We never understood why Mohamed always shrunk back from our sweet border collie, Addie, until I learned that dogs are unclean animals to Muslims. When I learned that the Qur'an is so holy to Muslims that many store the book in the highest place in their house, I remembered a time when he was showing me his and I grabbed it out of his hands to flip through it. When I was told it was brazen for a female to look directly into the eyes of a Muslim man, I thought about the times when I had innocently patted Mohamed's shoulder when he gave a correct answer in ESL class. It was probably amazing that he even came to the class to be taught by a woman at all. And yet here he was, our friend.

That was what was important: our beginning to know, understand, and befriend each other. When he was talking, Matt and I didn't point out his grammar mistakes. The verb tenses weren't the main point because we could understand what he was saying. And we found Mohamed giving us the same grace: when we made mistakes out of innocent ignorance, he leaned in anyway and befriended this strange American couple.

Timing with regard to locality and people is always interesting. As I said before, we met Mohamed in September 2001. I remember the ESL class that immediately followed the World Trade Center attacks. Everyone in the country was still in shock and mourning, and so was Mohamed. He denounced the attacks and expressed his sadness with short phrases like, "Very bad. Very bad." Over the next weeks and months, when "Let's Roll" turned into "Shock and Awe," it was hard for Matt and me to process it all, because, well, there was Mohamed. We couldn't just swallow photos of Muslim extremists from the news; we had a relationship with our friend Mohamed, and that muddied the waters in a good way.

Our friendship with Mohamed taught us many things, but I think one of the most important was this: that we should make sure, no matter our locality, to seek relationships with people that will show us different understandings of the world. In the same way that we don't (or shouldn't) read the

Bible just to reinforce our current understanding of Scripture, we shouldn't only be friends with people who reinforce our current understanding of life and the place in which we live it. Instead, we should let God lead us into spacious relationships.

Besides the immediate benefits of friendship with Mohamed, we found that knowing one person helped us to know a bit more about others like him. We would notice someone at the mall who dressed like him. We began to understand his religious calendar, enough so that when we were driving home and noticed a Muslim party at a nearby business, we knew they were celebrating Eid.

These kinds of relationships change things. There aren't quick judgments and lines aren't so easily drawn, because you recognize your common humanity. When you connect locality to people, you might become the annoying one who interjects, "Actually, that's not *completely* true . . ." when you hear generalizations about people and places. You can't help it, though, because the friendship you've experienced demands it. John Inge puts it like this: "Any conception of place is inseparable from the relationships that are associated with it."[2] You can't, as it says in Acts 4, help speaking about the things you've seen and heard. When you've seen the ways God works in different cultures among different peoples, you may have to be a missionary back to your own Christian culture.

Following Jesus means that we are to be the presence of love in the particular place where he's put us, among the people with whom we've been placed. Scripture is filled with instructions about engaging the people in our locality; we learn from the Old Testament that we are to love our neighbors as ourselves, and in the New Testament we hear of the Word that became flesh and blood and moved into the neighborhood. The word "neighbor" and its various forms are used 145 times in the New Testament alone.

Why do we read so much about this? If we see the Bible as a story of communication between God and people, we realize that the reason we read so much about it is because God spends so much of his time loving us. God's love for *everyone* characterizes the whole biblical story. First John 4:19, 21 says, "We, though, are going to love—love and be loved. First we were loved, now we love. He loved us first. . . . The command we have from Christ is blunt: Loving God includes loving people. You have to love both."

Jesus says in Matthew 5:45, "He gives his best—the sun to warm and the rain to nourish—to everyone, regardless: the good and bad, the nice and nasty." What he's saying is, if we want to love like God, we'll exhibit unconditional love for others within the context of relationship, place, and people. This is such a freeing statement. We get to love everyone! This means we don't have to spend time and energy figuring out who are the right ones to love. We're used to doing this, aren't we? We spend time and money trying to find out what is correct: What are the *right* shoes to go with this dress? Who are the *right* people to hang out with my kids? What are the *right* political beliefs? We try to size up people and places and determine whether or not we're going to wade into them.

Once, a man who was concerned about eternal life asked Jesus who were the right people to love. Jesus answered with a story about religious people who avoided a dying man in order to maintain their laws of purity. In the meantime, a good cultural and religious outsider saw himself in the dying man, reached out to him, and helped him. The moral of the story is not to spend so much time worrying about right and wrong. As the story says, we don't have a full understanding of God's idea of right and wrong, anyway. Instead, Jesus counsels us to love and connect to each other in our pursuit of eternal life together.

When we were newlyweds, a friend passed along this bit of wisdom in the form of a question: "Do you want to be right or do you want to be married?" And the question we should ask ourselves as people who follow Christ is the same: Are we more concerned about being right or about being in right relationship with people? Jesus summarizes his discussion of loving others like this in Matthew 5:48: "In a word, what I'm saying is, *Grow up*. You're kingdom subjects. Now live like it. Live out your God-created identity. Live generously and graciously toward others, the way God lives toward you."

When we're freed from the idea of being right, we get to rejoice in the fact that God allows us to love everyone. Shouldn't this thrill us? "Yes!" we should cheer. "We get to love *everyone!*" We Christ-followers have the opportunity to be the most open-minded, open-hearted people on the planet, thanks to his power in our lives. "Watch what God does, and then you do it," says the writer of Ephesians 5:1-2, "like children who learn proper behavior from their parents. Mostly what God does is love you. Keep company with him and learn a life of love. Observe how Christ loved us. His love was not cautious but extravagant. He didn't love in order to get something from us but to give everything of himself to us. Love like that."

We know how God loves us and we know God allows us to love all others. He asks us to give and receive love in wider and wider circles that revolve around his love at the core. Are we excited to see how wide the ripple effects of his love are? Are we hungering to see how God is working in the lives of those who are different from us?

God frees us not only from the need to be right but also from the need to see results. Many times we look to results to validate our efforts at loving others; if it was the right thing to do, we think, then we will be able to tell by the outcome. If we look to God for our model, once again we will see that we are freed from this restriction. "But God put his love on the line for us," writes Paul in Romans 5:8, "by offering his Son in sacrificial death while we were of no use whatever to him." If God loved us before we followed him, our anticipatory love for others is important no matter the results, because it goes before us. Being the presence of love in a place, says this Scripture, is the correct posture no matter the outcome. It's how we're the most like God.

And it's what sets us right with God. Paul writes this in Romans 13:8-10:

> When you love others, you complete what the law has been after all along. The law code—don't sleep with another person's spouse, don't take someone's life, don't take what isn't yours, don't always be wanting what you don't have, and any other "don't" you can think of—finally adds up to this: Love other people as well as you do yourself. You can't go wrong when you love others. When you add up everything in the law code, the sum total is *love*.

We can begin in our own contexts. We can start by making it a priority to have a close relationship with at least one person from another culture. How do you do this? Make phone calls, schedule meetings or chats over cups of coffee. If you are surrounded by people who are just like you, then go to where others are: shop at their markets, eat at their restaurants, walk in their neighborhoods. Most people, especially those transitioning to a new culture, are honored you are seeking them out.

We start by loving, by making ourselves available and being listeners. Find out what their needs are and find ways to provide help. "Each one of us needs to look after the good of the people around us, asking ourselves, 'How can I help?'" reads Romans 15:2-3. "That's exactly what Jesus did. He didn't make it easy for himself by avoiding people's troubles, but waded right in and helped out."

If we mean what we say about believing God, then his love is constantly changing us. He is consistently opening us up with his love so that we are enabled to love others. If this is happening, we should be able to look back at our lives, over the past five years and over the past five weeks, to see the results of God's work. We are in a new, spacious place. We can point to the love that we now have for people we once didn't care about. Can you do that? Can you point to the fruit of his love in your life? Who is in your life that you once didn't know but now desperately love? Let's go boldly into our own contexts and see the difference the love of God makes in the places and relationships in which he's put us.

Once I attended a lecture by a Mennonite author who had organized and participated in a campaign in which volunteers would simply ride buses. These buses happened to be in the West Bank, where bus bombs repeatedly decimated civilian lives. The idea behind the campaign, however, was that the presence of the love of God would be able to intervene, and the volunteer riders daily staked their lives on it.

Do we do this? Do we really believe God's love is that powerful? Do we know in our lives the game-changing power of his love—so much so that we are willing to get on the bus and ride with those who might be in danger?

Let's decide to love. Let's let go of our need to be right, our need to justify the results. Let's love everyone so freely that we're a curiosity to others. And when they ask us why, we can simply tell them we are following the example of a God who loves all people, including us.

Discussion Questions

1. Do your friendships reinforce your current understanding of life, or do they show you different views of the world?

2. Does my desire to be right or to see results get in the way of actively loving others?

3. Can you point to the fruit of God's love in your live? Whom in your life did you once not know but now desperately love?

Notes

1. John Inge, *A Christian Theology of Place* (Hampshire, Great Britain: Ashgate Publishing Ltd., 2003) 16.

2. Ibid., 26.

Chapter 15

Spacious God-Country

Why, the skies—the entire cosmos!—can't begin to contain him.[1]

Recently I was swimming with my family at an outdoor pool when I noticed that most of the people in the pool were looking up at the sky. I followed their gazes to a plane that flew skillfully back and forth, writing words with smoke. I don't know anything about flying or skywriting, but it seemed the perfect day for it; whatever message we were about to read would be the only sight in an endless blue sky. As the white block letters slowly emerged on the blue canvas, everyone around us paused to watch. First L . . . then O . . . then V . . . and finishing off with E, which was fun to watch as the plane twirled around to make the three parallel lines. It seemed as if all of us at the pool were thinking, "Yes! That's the perfect message to read in the sky on a sunny Sunday morning." As we went back to our splashing, you could hear people pointing out the word to friends and family. "Cool!" "Wow!"

But the show wasn't over yet. The pilot, after a few minutes, began writing again. This word was a little distant from the first, as if the pilot were starting a new thought. In fact, LOVE was beginning to fade as the next word leisurely appeared. We all watched again as the pilot stumped us with the first letter. We thought it was a C until he made a terrific loop around and added a small line to complete a G. He gradually finished the word with a couple more letters until two separate, puffy words hung in the lazy morning sky: LOVE and GOD. It was simply artful—worshipful, even.

As I admired the pilot's work, I heard a woman near me say to her friend, "Oh, it's a *religious* thing." It saddened me a little, the progression of this person's emotions. How is it that as the word LOVE hung in the sky, she was satisfied, possibly even inspired, and yet, when the pilot draped the word GOD alongside it, the meaning in the message was constricted? Doesn't the word GOD contain the same amount of wonder, mystery, and possibility? I didn't see the two words as mutually exclusive; I saw them as informing each other and defining each other. Yet, when the second word appeared, the message completely shut down for this woman. Why? I wondered. And, as a person of faith, I couldn't help wondering what we, the church, might have done in this woman's life to constrict the hope of the two words that hung there in the sky.

Is the thought of God—for this woman or for us—a confining or restrictive thought? When the average person on the street thinks of God, do images of hope and endless possibilities enter his mind? Or, at the mention of God, do the walls start slowly closing in on this person? Does he think he somehow has to find a way to be good before he is crushed?

Somehow, somewhere along the line, this poolside woman received the message that GOD was confining and restrictive. Did she get that message from us, the people who say we know and follow him? I guess the answer depends on how we follow him—how we talk about him, how we talk to him, and how we "wear" our faith. Ultimately, if we communicate to others that life with God is a pedantic, hair-splitting effort to hit a target, it might mean that *we* believe that's all it really is.

Nothing could be farther from the truth, according to Scripture. The God we read about in the Bible is characterized by mystery, wonder, and open-endedness. He's not preoccupied with Christian conformity; instead, we are told about a God who is so vast that we can spend our whole lives in relationship with him and still only glimpse his image. Life with this God is as limitless and immeasurable as the blue sky in which I read the words "LOVE" and "GOD."

We see this throughout Scripture, but I have recently noticed this spacious view of God in the story and writings of David. This shepherd-turned-king constantly spoke about the possibilities and the freshness his relationship with God brought to his life. It makes sense when we read his story; someone who spent so much of his life hiding out in caves or in exile would certainly use spatial imagery to describe peace. David constantly found himself in that room where the walls slowly close in. He faced a giant, a bitter has-been of a king, and feisty Philistine neighbors. And yet we read of someone who, even while backed into the stifling darkness of a cave, felt *freedom* in God. David, with stalactites dripping on him, composed poetry like this:

> He brought me out into a spacious place;
> he rescued me because he delighted in me. (2 Samuel 22:20 and Psalm 18:19, NIV)

> You gave me a wide place for my steps under me,
> and my feet did not slip. (Psalm 18:36, NRSV)
>
> Light, space, zest—that's GOD!
> So, with him on my side I'm fearless,
> afraid of no one and nothing. (Psalm 27:1, MSG)
>
> You didn't leave me in their clutches
> but gave me room to breathe. (Psalm 31:8, MSG)
>
> The spacious, free life is from GOD,
> it's also protected and safe.
> GOD-strengthened, we're delivered from evil—
> when we run to him, he saves us. (Psalm 37:39-40, MSG)
>
> Don't dump me, GOD;
> my God, don't stand me up.
> Hurry up and help me;
> I want some wide-open space in my life. (Psalm 38:21-22, MSG)
>
> I'm hurt and in pain;
> Give me space for healing, and mountain air. (Psalm 69:29, MSG)
>
> Do what you do so well:
> get me out of this mess and up on my feet.
> Put your ear to the ground and listen,
> give me space for salvation. (Psalm 71:2, MSG)
>
> Out of my distress I called on the LORD;
> the LORD answered me and set me in a broad place. (Psalm 118:5, NRSV)

Don't you want that kind of spaciousness in your life? I do. I want to be able to sit in the cave, as David did, and feel as if I'm standing in a meadow. I want to be able to face the battles in my life not with an asthmatic faith that constricts with circumstance but with God-given gulps of breathing room. What's more, I want that for others. I want to communicate to my poolside friend, and to others, that a life with God is about *real life*.

Life with God is a spacious, meaningful life. Is it a life filled with whatever you want whenever you want it? No. It's a life that attentively follows where God leads. "Obsession with self in these matters is a dead end," says Romans 8:6. "Attention to God leads us out into the open, into a spacious, free life." Is life with God a lazy, haphazard existence? No, it's a life where structure provides freedom. "An undisciplined, self-willed life is puny," we read in Proverbs 15:32. "An obedient, God-willed life is spacious."

God is the place where David ran when he needed wide-open spaces. All the political power or military might he had didn't provide him with the true peace he sought. When he went to God, when he talked and listened to God, David found exactly what he needed. If we listen, he is whispering to us in his poetry that he is set right only in God. "So spacious is he, so roomy," echoes Colossians 1:19, "that everything of God finds its proper place in him without crowding."

So where did this idea of a restrictive, controllable God come from? My guess is that it came from us, the humans. It seems we tend toward the tangible; we like objects we can carry around, things we can point to, measure, and purchase. Just the other day as I took a shower, I read the back of the bottle of body wash, which said this product "stimulates the soul and sharpens the senses." *Really?* I thought. If I need a boost for my soul, I just pop the top and squeeze? If only it were that easy! Another time, as I folded laundry, I was delighted to find the words "Knowledge, Wisdom, and Truth" printed inside the well-worn waistband of a pair of my pajama pants. How wonderful! I thought. We've answered Job's age-old question, "Where can wisdom be found?" Apparently, in this human age, it's no great mystery: truth is as near as a comfy pair of pajama pants.

Whether it's wisdom, love, or help for our souls, it seems we've stopped going to God and tried to provide love, wisdom, and truth for ourselves, or at least *the idea* of these things. We humans limit and confine God; God is the author of spacious places. He is not an object that we possess or something we squeeze out of a tube in order to make ourselves feel better.

This is not a new phenomenon. We read in Isaiah 44 of a similar problem. Even then, people limited their understanding of God by trusting in tangible things like wood and metal. And the truth that God explained to them is still applicable now: you limit God by confining who he is to objects. If you trust in limited, human things to sustain you, it only follows that you have a limited view of God.

But Holly, you might say, I'm not cooking my meals with firewood and then whittling out a carved deity, so I'm good. We don't tend to bow down to objects much anymore and pray to them, but do we trust in them just the same? We don't kneel before a copy of the *Wall Street Journal*, but are we trusting in the market to send our kids to college? We might not worship a particular career choice, but do we rely on our résumés to ensure we get to where we want to be?

Why do we do that? Isaiah 44:10 says, "Who would bother making gods that can't do anything, that can't '*god*'?" When we rely on finite things instead of the Infinite, we pull the drawstring around the spacious reality of the God who wants to lead us into wide-open life. We trust in our weapons to somehow get us out of the back of the cave instead of living into the freedom God provides us while we're sitting there in the dark.

What does this mean practically, all this sky and cave and pajama-pants talk? It means that we—*I*—need to live into the wide-open nature of who God is. We need to stop presuming upon him and instead begin watching him to see what surprising work of love he'll do next. We need to experience for ourselves and communicate to others the freedom and meaningful nature of life that comes with following Christ. We need to quit our fence-building and compartmentalizing and instead begin exploring this wild, spacious God-country. Let's trust in his ability to set the proper boundaries and not our own human tendency to confine and objectify. Let's embody the fact that LOVE and GOD don't bind up but instead release.

There's a wideness in God's mercy, says one old hymn, and I tend to agree:

> There's a wideness in God's mercy,
> Like the wideness of the sea;
> There's a kindness in his justice,
> Which is more than liberty.
>
> There is welcome in God's mercy,
> And more graces for the good;
> There is mercy with the Savior;
> There is healing in his blood.
>
> But we make his love too narrow
> By false limits of our own;
> And we magnify his strictness
> With a zeal he will not own.
>
> For the love of God is broader
> Than the measure of man's mind;
> And the heart of the Eternal
> Is most wonderfully kind.[2]

Isn't *this* the God we know? Isn't *this* the God we should reflect?

In the Disney/Pixar movie *Up*, we meet a girl named Ellie whose dreams and escapades inspire the rest of the movie's characters and plot. Her favorite saying, one we hear her squeaky voice say numerous times in the film, is that of her favorite explorer: "Ad-VEN-ture is OUT THERE!" And while it may seem like an obvious statement to us, at some point we have to come to that conclusion ourselves. In some way, we men and women have to come to the end of ourselves and realize that true life is not found inside us or within something we've created. Adventure is *out there*. This true life in Christ may have boundaries and structure, but it is a life that leads us *out there*, beyond ourselves, to spaciousness and freedom.

Let's not only live into this adventurous reality ourselves but also point the way for those who don't yet see how LOVE and GOD could hang in the same blue sky. "God's spirit beckons," Paul once wrote. "There are things to do and places to go!"[3] Skip the body wash, he says, and leave behind the smallness of who you now think God is. Go, as David did, to the spacious place in the heart of God. Maybe then you, too, will be able to live out the poetry of "He makes me lie down in green pastures, he leads me beside quiet waters, he restores my soul."

Discussion Questions

1. Where do you see evidence of our human tendency to confine and objectify God?

2. What Scriptures about the spacious nature of God caught your attention? Why?

Notes

1. 2 Chronicles 2:6.

2. Frederick W. Faber, "There's a Wideness in God's Mercy," *The Baptist Hymnal* (Nashville: Convention Press, 1975) 171.

3. Romans 8:14.

Chapter 16

Ripping and Sewing

God is not in a protected place shrouded in a hermetically sealed ontological box. God is, rather, a shocking presence in a word of ambiguities.[1]

I am sitting in the midst of a mess, trying hard not to clean it. As I type in our small living room, I can look around at half-completed paper crafts, a bag of 100 gumdrops ready for the 100th day of school party (which is next week), winter coats from the bus stop run, souvenir tickets from a recent football game, a broken night light, and tiny wood chips on the carpet in front of the fireplace. Let's not get started on the kitchen or the baskets of laundry waiting for me in the basement.

It's hard for me to relax and work in this setting. Call me a Chihuahua if you will, but I have the impulse to use thirty minutes or an hour to carry each of these things to its proper place. *Then* I can write, I think. The problem is, if I take an hour of writing time here and another one there to clean (all of which will be obliterated in the five minutes everyone walks in the door), when do I actually get my work done?

So I am trying hard to sit and concentrate amid the mess. After all, who says good writing only happens when you're sitting in an office straight out of a Pottery Barn catalog? There will be time for cleaning—and more hands to help me—later. So maybe this isn't what Virginia Woolf had in mind when she was talking about having a room of one's own; it's just the place I have today to get some writing done.

It occurs to me that this attitude carries over into my spiritual awareness. Can I give myself and others grace in the middle of a messy situation? Can I move forward in the midst of a not-quite-ideal relationship? Am I able to be fully present while holding difficult aspects of my life in tension, or does the "mess" irritate and distract me?

Author John Fischer says this inability to give and receive grace can sometimes carry over into our view of God, something he calls being a utopic Christian. "Utopic Christians can't see God in the flawed, in the disappointment, in the poor, or in the unfinished quality of their lives—even in the average. They see him in the winners, not the losers. They see him in victory, not defeat."[2]

I'm not talking about accepting the status quo or wandering through life with no goals or accountability. I'm talking about my need to keep before me the fact that there is process in life. Most important things, like family and home, take time and create messes. There has to be a place for do-overs; we have to allow for growth. My mom would say, usually when she saw me use the seam ripper to correct a mistake in something I was making, "Ripping is a part of sewing." Maybe I should consider getting that tattooed on my body somewhere. I'm surprised at how ungracious I am at times and how quickly I get annoyed with the "mess" in life. When did I become so intolerant of the messy? After all, *I'm* messy. I constantly say the wrong words and do the wrong things. My insecurities get the best of me, and when life doesn't look like the Pottery Barn life, I immediately begin to ask myself what the issue is or what I could have done differently.

God is clearing more space for grace in my life. I'm working to develop the constant awareness that everyone is learning and we're all on a journey together. It's a process. It's not easy, especially when I'm in the middle of conflict. I'm trying to follow him through it, though, and see conflict as a potential opportunity for grace and growth.

God is the perfect one to develop this in me; when he talks about himself in Scripture and reveals himself to us in Christ, he uses the word "grace" to define himself. When Moses asks for a closer relationship with God and asks him to teach him his ways, God comes to him on Mount Sinai, and for the first time he defines himself. "GOD, GOD, a God of mercy and grace, endlessly patient," he calls out as he comes to Moses. "So much love, so deeply true, loyal in love for a thousand generations, forgiving iniquity, rebellion, and sin. Still, he doesn't ignore sin." This is important because we can see how God explains himself as gracious and patient. "Nowhere before this speech has anyone been privileged to hear directly a disclosure of what is most powerful and definitional for God's own life."[3] This is also notable because the conversation between God and Moses is taking place after the infamous golden calf episode. This is not, then, just a parent coming to kiss his children goodnight. It is a parent kissing his children goodnight at the end of a knock-down-drag-out day.[4]

This episode would have been in the minds of the readers of John 1. To us, the connection may not be obvious, but the ancient audience would have heard this definition of God's merciful character echoed in the grace-and-truth wording of John 1:14, almost as if he were quoting it: "The Word became flesh and made his dwelling among us. We have seen his glory, the glory of the One and Only, who came from the Father, full of grace and truth."[5] John was saying, "Remember the God from Mount Sinai? The one who is just and yet still forgives? Jesus *is* this God. This same grace and truth is found in *him*."[6]

We read of this gracious and truthful character of God yet again in 2 Peter 3:9. The New Revised Standard Version says that God is "patient with you, not wanting any to perish, but all to come to repentance." *The Message* paraphrase says that God "doesn't want anyone lost. He's giving everyone space and time to change." This doesn't mean we get to lollygag and only follow Christ when we get around to it. It is a spacious declaration of the time our journey takes. It is a call to give that freeing gift to others because we recognize we are all growing.

I have never been what you would call a coordinated person. My life is littered with stories of my dropping breakables, walking into cabinet corners, and tripping over the painted lines on sidewalks. Once, a girl on my high school soccer team asked me, "Why is it that every time you get the ball on a breakaway and *nobody* is around you, you trip and fall?" It was true; I'd get so excited when it was just me on the goalie, it was almost as if I would slide-tackle myself.

I've learned to cope with my clumsiness over the years, and my husband has learned to have ice packs and heating pads ready. My daughter seems to have gotten her father's body-awareness gene because she spends her time doing handstands and using a Hula-Hoop with no major mishaps. At first, Mikias seemed to avoid any environmental side effects of having an awkward mommy as well, running around and swimming with the rest of us.

Then came last summer. Mikias was three and a sight to see. He tripped and fell so many times that his knees, uncovered in shorts, were a bloody mess from May until September. We couldn't have enough bandages on hand as the same scabs kept opening up. It wasn't just scraping his knees; it was knocking over cups at the dinner table. His record was four drink spills in one meal (one of which we were sharing with a couple who has yet to have kids). None of these things were his fault, but what was going on?

Growth spurts, a friend said, and I was intrigued. Mikias's body was growing at an enormous rate, and the sensory pathways were racing to keep up. His arms and legs were longer before his brain knew it. His center of gravity had sneakily shifted, almost like the floor in a funhouse. Some sources say kids can grow as much as one centimeter in a twenty-four-hour period, affecting the body like an implosion.[7]

What if we gave ourselves grace in the midst of our growth spurts? God does. How would our lives be different if we came to others with bandages ready because we know they're in the process of growing? That's what Christ did. What if we armed ourselves with paper towels, ready to mop up yet another spill in the life of a friend whose has parts of himself that haven't caught up to the others yet? We might even need to explain it to him so that he doesn't feel guilty or awkward: "You're growing right now. Sometimes falls and spills happen. Just get back up because, eventually, you'll get it."

God doesn't ask us to have it all together before he starts working in our lives, just as he didn't demand the perfection of the world before he entered it. I love the way Barbara Robinson depicted this idea in her 1972 book, *The Best Christmas Pageant Ever*. What happens when the Herdmans, the meanest kids in town, tumble into the church and take over all the lead parts in the annual Christmas program? The church people can't handle the mess. They get irritated because their view of the holy family is changed when the roles are played by skinny-legged kids who argue, throw punches, and smoke cigars. The main character is a young girl caught between Alice, her church friend who has always gotten to be Mary, and Imogene Herdman, who bullies her way into playing the mother of Christ. On the night of the pageant, this is what happens as the Herdmans enter the sanctuary:

> Ralph and Imogene were there all right, only for once they didn't come through the door pushing each other out of the way. They just stood there for a minute as if they weren't sure they were in the right place—because of the candles, I guess, and the church being full of people. They looked like the people you see in the six o'clock news—refugees, sent to wait in some strange ugly place, with all their boxes and sacks around them.
>
> It suddenly occurred to me that this was just the way it must have been for the real Holy Family, stuck away in a barn by people who didn't much care what happened to them. They couldn't have been very neat and tidy either, but more like *this* Mary and Joseph (Imogene's veil was cockeyed as usual, and Ralph's hair stuck out all around his ears). Imogene had the baby doll but she wasn't carrying it the way she was supposed to,

cradled in her arms. She had it slung up over her shoulder, and before she put it in the manger she thumped it twice on the back.

I heard Alice gasp and she poked me. "I don't think it's very nice to burp the baby Jesus," she whispered, "as if he had colic." Then she poked me again. "Do you suppose he could have had colic?"

I said, "I don't know why not," and I didn't. He *could* have had colic, or been fussy, or hungry like any other baby. After all, that was the whole point of Jesus—that he didn't come down on a cloud like something out of "Amazing Comics," but that he was born and lived . . . a real person.[8]

So many times, we want to be the person who is the hero, the one who patiently lets the Herdmans into the pageant so they can begin to grasp the grace of the Christmas story. What that means, however, is that I'm going to have to be okay with the mess. I'm going to have to treat others gently as I deal with their "stuff" and remember that they have to put up with my "stuff" as well.

In the grand scheme of things, the idea that none of us have "stuff" or sin in our lives is a myth. We are all constantly in need of forgiveness, and thank God that he is willing to engage us, bandages ready. Eugene Peterson writes, "God's great love and purposes for us are worked out in the messes, storms and sins, blue skies, daily work, and dreams of our common lives, working with us as we are and not as we should be."[9] He is willing to teach us not a list of regulations that will make us perfect and acceptable to him but a way of interacting and living with him that allows us to be who he created us to be. "Walk with me and work with me," says Jesus in Matthew 11:29-30. "Watch how I do it. Learn the unforced rhythms of grace. I won't lay anything heavy or ill-fitting on you. Keep company with me and you'll learn to live freely and lightly."

Let us be grateful that we have a Father who gives us all space and time to change. May we who follow Christ recognize others' rights to the same space and time, and rush to give it freely, with bandages ready.

And the next time you're at my house and it's a real mess, please congratulate me.

Discussion Questions

1. What does it mean to you that God gives everyone space and time to change?

2. Ask God to teach you his unforced rhythms of grace.

Notes

1. Philip Sheldrake, *Spaces for the Sacred* (Baltimore MD: Johns Hopkins University Press, 2001) 67.

2. John Fischer, *Finding God Where You Least Expect Him* (Eugene OR: Harvest House, 2003) 15–16.

3. Walter Brueggemann, *Exodus*, The New Interpreter's Bible, vol. 1 (Nashville: Abingdon Press, 1994) 946.

4. Exodus 32–34.

5. This passage is from the NIV.

6. Gail R. O'Day says, "'Full of grace and truth' echoes the Hebrew word pair 'steadfast love' and 'truth' (*hesed* and *emet*; e.g., Exod 34:6) that speaks of God's covenantal love and faithfulness. The concentration of OT languages and imagery intensifies as the focus of the Prologue turns from the eternal Word to the incarnation." For more, see Gail R. O'Day, John, The New Interpreter's Bible, vol. 9 (Nashville: Abingdon Press, 1995) 523.

7. http://life.familyeducation.com/teen/puberty/42924.html#ixzz1kIyGVLuh (accessed on 23 January 2012).

8. Barbara Robinson, *The Best Christmas Pageant Ever* (New York: Scholastic, 1972) 73–74.

9. Eugene H. Peterson, "Introduction to Joshua," *The Message* (Colorado Springs: NavPress, 2003) 232.

Chapter 17

From the Creative Place

What I do is me: for that I came.
—Gerard Manley Hopkins[1]

One of the things I enjoy most when visiting a new place is seeing the way biblical stories are represented visually. Museums, cathedrals, and markets will surprise you with thought-provoking interpretations of stories you've heard over and over. A drawing on the wall might catch your eye, and you see a man with traditional Zulu hair and clothes reaching out to an AIDS patient. All of a sudden, you see nail marks in the hands of the Zulu caretaker, and a wound in his side. Wait a minute, you say. I know that story. And though it doesn't look exactly like the mental image you've had in your head (the one that might've been put there by a flannel graph or a Precious Moments Bible somewhere along your journey), you connect with the truth the artist is communicating.

Maybe you are in a market and you see a handmade pillowcase that someone is selling. It's a simple, blue square with a giant gray fish embroidered on it with coarse thread. And since you don't have a marine-themed living room, you move on. But wait. What is embroidered in the fish's mouth? It's a small man, entering the mouth of the great fish. Hold on, you think. Now I understand. And you pick up the pillowcase and pay for it because you love the way the seamstress has captured the hope and the fear in the story.

Whatever the culture, whatever the medium, it is exciting to see the ways that people all over the world communicate these stories and their life-changing truths. The act of creation is miraculous; truth is communicated through a particular person of a particular skill from within a particular culture and time in history. We could read facts about this artist's life or about the story he is trying to tell, but it wouldn't communicate on the level that the canvas or the piece of carved wood does. And we realize through their creativity that, though we could take a lifetime enjoying and studying the artwork inspired by the biblical account, these must surely pale in comparison to the ongoing imaginative works of the God in whom the creative act originated.

God's medium is not just the chaos at the beginning of time, and his methods are not confined to a seven-day span. God artfully continues to craft, right up to this very day. And while we see the evidence in seedlings and spring showers, his creativity comes into its own in the lives of people. "My Father is always at his work to this very day," Jesus once said, "and I, too, am working."[2] God is at work, creating and restoring. This imaginative process involves individuals as well as groups, cities, and nations. And just as we read in Matthew 6, the beauty of the lilies is outdone by the beauty God works in the lives of people.

Don't you want to see it? Are you on the watch for it? Like one who rises early in order to take in the sunrise, I want to seek the creative work of God with intention. I want to be on the lookout for his molding and shaping in my own life as his spirit leaves its fingerprints on me. I want to go to places where I can see the artful ways Christ is setting things right all over the world. I need to take in the creative interpretations of Christ's work in others, for it inspires change in me as well. I want to celebrate the diverse methods God uses. I long for the salvation images of other cultures to color my understanding of Scripture. There is such joy in this discovery. There are countless reasons to celebrate God's ongoing creative process.

God's creative action is spacious. As we saw in previous chapters, God uses particular and diverse cultural forms to help us grasp his grace and love. The beauty of his work is only enhanced by the variety of methods he uses to create rightness in the world. "For Christ plays in ten thousand places," writes poet Gerard Manley Hopkins, "Lovely in limbs, and lovely in eyes not his / To the Father through the features of men's faces."[3]

Paul greatly prizes the Spirit's creativity, and we read a poignant argument for freedom in faith within his letter to the Galatian church. The church, he discovers, is being swayed by well-meaning people who have positioned themselves as theology gatekeepers, deciding what is orthodox or religiously appropriate when it comes to following Christ. The prominent cultural form being discussed is the necessity of circumcision for people who are following Jesus but aren't culturally Jewish. Your job, Paul emphatically writes to the church, is to watch for and embrace God's creative action, not to ensure homogeneity among members. Paul explains to the church that circumcision is a cultural and religious form, something that represents an idea of faith to a particular people. And while the ideas of grace and covenant remain consistent, Paul is adamant that God can artfully use what or whom he chooses in communicating his message of love. We read in Acts and in Galatians that neither Peter nor Paul dreams up these ideas of freedom of their own accord; they are simply observers of God's creative action. When they see the same Spirit at work in the lives of Gentiles, they celebrate the ability of God to make the world even more whole.

It is easy, though, as we've explored in other chapters, to begin to codify the way we interact with God. Our tendency as humans is to take the particulars of our own faith experience and call them "right." We take our religious forms and turn them into a system of regulations. We try to communicate our choices in well-meant ways that eventually look like a long set of rules. And, while structure does provide freedom as we follow Christ, the books of Acts and Galatians are strong warnings to the church that we are not to be preoccupied with Christian conformity. We are not the monitors of what is orthodox or what isn't; God is. As we've discussed before, God is going before. We're late to the meeting. We are watching his artistic process, and sometimes we are lucky enough to participate. Our view of God's work is always limited, always incomplete.

What if we, as people who follow Christ, committed ourselves to be willing, open, and excited about what God is doing? What if we devoted ourselves to cheering on the new and different ways He is drawing people to himself? What if we meditated on the biblical examples of people with this openness? There are many: Simeon in the temple; Ananias who receives the blinded Saul; Barnabas, the Son of Encouragement. We read story after story about people who simply trust and take joy in the ways they see God working. These are not people who are caught up in what they should or shouldn't do.

Paul reminds the church at Galatia, and us, that we shouldn't be caught up in it either. "Rule-keeping does not naturally evolve into living by faith," he says in 3:12, "but only perpetuates itself in more and more rule-keeping." In 3:22, he writes, "if any kind of rule-keeping had power to create life in us, we would certainly have gotten it by this time." And further in 3:11, "The obvious impossibility of carrying out such a moral program should make it plain that no one can sustain a relationship with God that way. *The person who lives in right relationship with God does it by embracing what God arranges for him.* Doing things for God is the opposite of *entering into what God does for you.*"[4] What a powerful testimony Paul gives to the spacious life Christ creates; we enter obediently into the beauty of what the Creator makes for us.

Could it be that we have a deficient theology of the way the Holy Spirit works in our lives and in the lives of others? Is it that we don't trust the Spirit enough to lead a person in the way he or she should follow Christ? "We are *all* able to receive God's life, his Spirit, in and with us by believing," Paul writes in Galatians 3:14. What if we Christians were to relinquish our "positions" as the Sunday school class monitors and instead look to the Holy Spirit as he continues to create his grand works of art? Elsewhere, Paul writes, "Those who trust Christ's work in them find that God's Spirit is in them—living and breathing God!"[5]

Paul writes powerfully, challenging the church to live into the creative action of God instead of living into its own sense of religious identity. It's not about freedom for freedom's sake or about the ability to do only what we want or only what we know. Instead, Paul counsels us against that kind of selfish existence. That isn't freedom, he says. "When you attempt to live by your own religious plans and projects, you are cut off from Christ, you fall out of grace," he writes in Galatians 5:3-4. "I have been crucified in relation to the world," he continues in 6:14-15, "set free from the stifling atmosphere of pleasing others and fitting into the little patterns that they dictate. Can't you see the central issue in all this? It is not what you and I do—submit to circumcision, reject circumcision. It is what *God* is doing, and he is creating something totally new, a free life!"

In reading about adoption and attachment, you find many case studies of families where bonding doesn't ever fully occur. The saddest to me are the families in which an adopted child never quite feels like a member of the family. The child is afraid of being abandoned again, and so she tries to make sure not ever to cause trouble. Many times this fear exhibits itself through perfectionism: the child is afraid *not* to make her bed or afraid what will happen if she *doesn't* excel in school. I read once of a child who felt like she was living in a hotel her whole life instead of in a family home, always making sure her room was spotless. How sad! A family went to great lengths to show love to a child and, for some reason, that child still missed the love, freedom, and comfort of being in a family.

While I can't say that my family has adoption attachment all figured out, I imagine that this scenario might feel vaguely familiar to God. Does he look at our worship each Sunday and see us dutifully making our beds? Has he gone to great lengths to free us and call us his own only to find us cowering behind a list of rules? Do we think rules are all he cares about? Are we afraid if we mess up once, we're on our own again? "Christ has set us free to live a free life," writes Paul in Galatians 5:1, but do we, like the girl in the case study, sometimes miss the freedom in being made a part of family?

How can we fully live into the artful reality God is working in the world and in our lives? Paul writes in Galatians 5:16, "My counsel is this: Live freely, animated and motivated by God's Spirit." We must trust the Spirit working inside us; the same Spirit who hovered over the waters of chaos in the beginning wants to form a masterpiece from the chaos of our lives. "Thus we have been set free to experience our rightful heritage," says Paul in Galatians 4:5-7. "You can tell for sure that you are now fully adopted as his own children because God sent the Spirit of his Son into our lives crying out, 'Papa! Father!' Doesn't that privilege of intimate conversation with God make it plain that you are not a slave, but a child?"

But how do we continue? What are the steps? Isn't there an acronym for this kind of thing? Not really, Paul says. Instead, we find ourselves caught up in the ongoing, organic artistry of God's work in us. The ways of following Christ are as varied as the people who follow him. It should be a continuous, unpredictable process. "Continue to work out your salvation with fear and trembling," Paul writes elsewhere, "for it is God who works in you."[6]

Instead, we are to put our energies into discovering the effects of Christ's salvation work in us. It's a daily effort that affects us as we feed the cat, wait in the carpool, and work for the corporation. "Since this is the kind of life we have chosen, the life of the Spirit," Paul writes in Galatians 5:25 and 6:1, "let us make sure that we do not just hold it as an idea in our heads or a sentiment in our hearts, but work out its implications in every detail of our lives. . . . Each of us is an original. Live creatively."

We have the chance each day to *live from this creative place*, to uncover the effects of Christ's creative restoration. We have the opportunity to live freely following the Spirit's lead. It's our job to listen, to attentively observe, and then live fully into who God has made us. There is purpose in his artistry and we best not waste it. "Make a careful exploration of who you are and the work you have been given," writes Paul in Galatians 6:4-5, "and then sink yourself into that. Don't be impressed with yourself. Don't compare yourself with others. Each of you must take responsibility for doing the creative best you can with your own life." As we have been freed, we must also make sure to help others discover the divine originality God has worked in their own lives and in their own cultures.

A friend introduced me to a book titled *Ish* by Peter H. Reynolds. It tells of a boy named Ramon who loves to draw until one day his brother, Leon, bursts out laughing at his interpretation of a vase. All of a sudden, drawing isn't fun for Ramon. Since he can't get any of his pictures to look "right," he decides he is done with drawing altogether. He crumples up his vase sketch and throws it to the floor. Ramon's sister, Marisol, picks up the crumpled paper and declares it to be one of her favorites because it is "vase-ish." As Ramon begins to look at his drawings in this way, he is energized. Drawings spring from his pencil without worry, and drawings inspire more art, which inspire stories and then poems. At the end of the story, we find an artful Ramon living "ishfully ever after."[7]

Will we, as followers of Christ, be bold enough to embrace the freedom he has provided for us? Will we be consumed with living what we think is "right," or will we let the Spirit lead us? Maybe the harder question is, which of Ramon's siblings do we want to be? Will we squelch the interpretations of others even as we see the markings of God's Spirit in the work? Will we shrug off another's expression of his or her faith in Christ because it isn't how we would do it? Or will we pick up and smooth the work of those who might need encouragement to be exactly who God made them—and infuse life and energy into their faith?

May the God who is loving enough to create and enjoy the spectrum of humanity lead us all in creative, spacious freedom. Because the truth is, all of our depictions of faith are "faith-ish."

Discussion Questions

1. Where have you noticed God's ongoing creative action recently?

2. How do you embrace what God arranges for you? Are you doing your creative best with your life?

Notes

1. Hopkins's poem, "As kingfishers catch fire," is examined at length in Eugene H. Peterson, *Christ Plays in Ten Thousand Places* (Grand Rapids MI: Eerdmans, 2005) 1–2.

2. John 5:17, NIV.

3. Quoted in Peterson, *Christ Plays*, 1–2.

4, Italics mine.

5. Romans 8:5, MSG.

6. Philippians 2:12, NIV.

7. Peter H. Reynolds, *Ish* (Cambridge: Candlewick Press, 2004) 27.

Conclusion

One of the remarkable qualities of the story is that it creates space.
We can dwell in a story, walk around, find our own place.
The story confronts but does not oppress; the story inspires but does not manipulate.
The story invites us to an encounter, a dialog, a mutual sharing.[1]

When visiting other places, you often meet people whose given names are English words. I've met Happy, Pretty, Precious, and Blessed. I've met lots of Gifts and countless Princes and Princesses. I guess it's not that uncommon here in the United States as well: I have a lovely aunt named Faith, friends named Grace, and I once had a classmate named America. But the most powerful name I remember running across while abroad was one I saw on the name tag of a young African man: Listen.

Listen. It was a command, a request. It almost begged us to speak in a whisper, or not at all, because he might tell us something essential. Or maybe his name was a reminder, wise counsel passed down from his predecessors, for him and for us. The name had both a mystical and practical quality to it. It was a spiritual word and one we used every day. It was almost as if his name beckoned us on, beyond ourselves.

Listening, in one way or another, is what we've been exploring. Our world is full of organisms (plants, animals, people) that hear and need to be heard. This book is simply a compilation of reflections from one of those people trying to voice the connections I've observed between faith and place. It's filled with personal faith stories and place stories . . . and you've graciously listened.

It's your turn. Listen to God. Listen to your place. Listen to *God in your place*. Pause and hear the spaciousness in your everyday life. As you pause, remember the grace-filled promise we find in Scripture: when we go to the place of God, we will find a listener. We will find him present and at work in every one of our places.

"Can it be that God will actually move into our neighborhood?" asked King Solomon as he dedicated the temple.[2] And yet, in his prayer, he was bold enough to repeatedly ask that God listen to them.

Generations later, another king stepped out in the same audacious belief. Jehoshaphat of Judah received word that war as at the doorsteps of his kingdom. Badly shaken, he went to God for help and ordered a nationwide fast. Eventually, everyone in the kingdom was amassed as the king prayed.

> When the worst happens—and we take our place before this Temple (we know you are personally present in this place!) and pray out our pain and trouble, we know that you will listen and give victory.
>
> And now it's happened: men from Ammon, Moab, and Mount Seir have shown up. . . . O dear God, won't you take care of them? We're helpless before this vandal horde ready to attack us. We don't know what to do; we're looking to you.

And I love what the Bible says next: "Everyone in Judah was there—little children, wives, sons—all present and attentive to GOD."[3] The amazing act of deliverance that Jehoshaphat and his kingdom eventually see is proof of the resounding theme throughout Chronicles and the whole of Scripture: "If you seek him, he will be found by you."[4]

But before we rush to the victorious songs of Judah, remember my African friend's name and advice. Picture the whole nation of Judah—little children, wives, sons—standing before God. *The Message* says they were "present and attentive." What were they doing? They were listening . . . and so was God.

We all want to be listened to, and we need to listen. We need to hear, and at the same time we desperately want to be heard. "Faith comes from hearing," says one version of Romans 10:17, and another one says, "The point is, before you trust, you have to listen."[5]

In attempting to voice place and faith, I've simply learned that to follow God means to listen to him and to the landscape of my life. May we, the followers of Christ, listen. May we live present, attentive lives that walk in obedience into the ambiguous, moving forward in the knowledge that his spacious life is just up ahead.

Notes

1. Henri J.M. Nouwen, *The Living Reminder* (New York: Seabury Press, 1977) 66.
2. 2 Chronicles 6:18.
3. 2 Chronicles 20:8-13.
4. 2 Chronicles 15:2, NRSV.
5. The first translation of this verse is from the NIV and the second is from *The Message*.

Other available titles from

Beyond the American Dream
Millard Fuller

In 1968, Millard finished the story of his journey from pauper to millionaire to home builder. His wife, Linda, occasionally would ask him about getting it published, but Millard would reply, "Not now. I'm too busy." This is that story. 978-1-57312-563-5 272 pages/pb **$20.00**

The Black Church
Relevant or Irrelevant in the 21st Century?
Reginald F. Davis

The Black Church contends that a relevant church struggles to correct oppression, not maintain it. How can the black church focus on the liberation of the black community, thereby reclaiming the loyalty and respect of the black community? 978-1-57312-557-4 144 pages/pb **$15.00**

Blissful Affliction
The Ministry and Misery of Writing
Judson Edwards

Edwards draws from more than forty years of writing experience to explore why we use the written word to change lives and how to improve the writing craft. 978-1-57312-594-9 144 pages/pb **$15.00**

Bottom Line Beliefs
Twelve Doctrines All Christians Hold in Common (Sort of)
Michael B. Brown

Despite our differences, there are principles that are bedrock to the Christian faith. These are the subject of Michael Brown's *Bottom Line Beliefs*. 978-1-57312-520-8 112 pages/pb **$15.00**

Christian Civility in an Uncivil World
Mitch Carnell, ed.

When we encounter a Christian who thinks and believes differently, we often experience that difference as an attack on the principles upon which we have built our lives and as a betrayal to the faith. However, it is possible for Christians to retain their differences and yet unite in respect for each other. It is possible to love one another and at the same time retain our individual beliefs.

978-1-57312-537-6 160 pages/pb **$17.00**

To order call 1-800-747-3016 or visit www.helwys.com

Choosing Gratitude
Learning to Love the Life You Have
James A. Autry

Autry reminds us that gratitude is a choice, a spiritual—not social—process. He suggests that if we cultivate gratitude as a way of being, we may not change the world and its ills, but we can change our response to the world. If we fill our lives with moments of gratitude, we will indeed love the life we have. 978-1-57312-614-4 144 pages/pb **$15.00**

Contextualizing the Gospel
A Homiletic Commentary on 1 Corinthians
Brian L. Harbour

Harbour examines every part of Paul's letter, providing a rich resource for those who want to struggle with the difficult texts as well as the simple texts, who want to know how God's word—all of it—intersects with their lives today. 978-1-57312-589-5 240 pages/pb **$19.00**

Dance Lessons
Moving to the Beat of God's Heart
Jeanie Miley

Miley shares her joys and struggles a she learns to "dance" with the Spirit of the Living God. 978-1-57312-622-9 240 pages/pb **$19.00**

The Disturbing Galilean
Essays About Jesus
Malcolm Tolbert

In this captivating collection of essays, Dr. Malcolm Tolbert reflects on nearly two dozen stories taken largely from the Synoptic Gospels. Those stories range from Jesus' birth, temptation, teaching, anguish at Gethsemane, and crucifixion. 978-1-57312-530-7 140 pages/pb **$15.00**

Divorce Ministry
A Guidebook
Charles Qualls

This book shares with the reader the value of establishing a divorce recovery ministry while also offering practical insights on establishing your own unique church-affiliated program. Whether you are working individually with one divorced person or leading a large group, *Divorce Ministry: A Guidebook* provides helpful resources to guide you through the emotional and relational issues divorced people often encounter.
978-1-57312-588-8 156 pages/pb **$16.00**

To order call 1-800-747-3016 or visit www.helwys.com

The Enoch Factor
The Sacred Art of Knowing God
Steve McSwain

The Enoch Factor is a persuasive argument for a more enlightened religious dialogue in America, one that affirms the goals of all religions—guiding followers in self-awareness, finding serenity and happiness, and discovering what the author describes as "the sacred art of knowing God."
978-1-57312-556-7 256 pages/pb **$21.00**

Faith Postures
Cultivating Christian Mindfulness
Holly Sprink

Sprink guides readers through her own growing awareness of God's desire for relationship and of developing the emotional, physical, spiritual postures that enable us to learn to be still, to listen, to be mindful of the One outside ourselves.
1-978-57312-547-5 160 pages/pb **$16.00**

The Good News According to Jesus
A New Kind of Christianity for a New Kind of Christian
Chuck Queen

In *The Good News According to Jesus*, Chuck Queen contends that when we broaden our study of Jesus, the result is a richer, deeper, healthier, more relevant and holistic gospel, a Christianity that can transform this world into God's new world.
978-1-57312-528-4 216 pages/pb **$18.00**

Healing Our Hurts
Coping with Difficult Emotions
Daniel Bagby

In *Healing Our Hurts*, Daniel Bagby identifies and explains all the dynamics at play in these complex emotions. Offering practical biblical insights to these feelings, he interprets faith-based responses to separate overly religious piety from true, natural human emotion. This book helps us learn how to deal with life's difficult emotions in a redemptive and responsible way.
978-1-57312-613-7 144 pages/pb **$15.00**

Hope for the Thinking Christian
Seeking a Path of Faith through Everyday Life
Stephen Reese

Readers who want to confront their faith more directly, to think it through and be open to God in an individual, authentic, spiritual encounter will find a resonant voice in Stephen Reese.
978-1-57312-553-6 160 pages/pb **$16.00**

To order call 1-800-747-3016 or visit www.helwys.com

Hoping Liberia
Stories of Civil War from Africa's First Republic
John Michael Helms

Through historical narrative, theological ponderings, personal confession, and thoughtful questions, Helms immerses readers in a period of political turmoil and violence, a devastating civil war, and the immeasurable suffering experienced by the Liberian people.

978-1-57312-544-4 208 pages/pb **$18.00**

A Hungry Soul Desperate to Taste God's Grace
Honest Prayers for Life
Charles Qualls

Part of how we *see* God is determined by how we *listen* to God. There is so much noise and movement in the world that competes with images of God. This noise would drown out God's beckoning voice and distract us. We may not sense what spiritual directors refer to as the *thin place*—God come near. Charles Qualls's newest book offers readers prayers for that journey toward the meaning and mystery of God.

978-1-57312-648-9 152 pages/pb **$14.00**

James (Smyth & Helwys Annual Bible Study series)
Being Right in a Wrong World
Michael D. McCullar

Unlike Paul, who wrote primarily to congregations defined by Gentile believers, James wrote to a dispersed and persecuted fellowship of Hebrew Christians who would soon endure even more difficulty in the coming years.

Teaching Guide 1-57312-604-5 160 pages/ pb **$14.00**
Study Guide 1-57312-605-2 96 pages/pb **$6.00**

James M. Dunn and Soul Freedom
Aaron Douglas Weaver

James Milton Dunn, over the last fifty years, has been the most aggressive Baptist proponent for religious liberty in the United States. Soul freedom—voluntary, uncoerced faith and an unfettered individual conscience before God—is the basis of his understanding of church-state separation and the historic Baptist basis of religious liberty.

978-1-57312-590-1 224 pages/pb **$18.00**

To order call 1-800-747-3016 or visit www.helwys.com

The Jesus Tribe
Following Christ in the Land of the Empire
Ronnie McBrayer

The Jesus Tribe fleshes out the implications, possibilities, contradictions, and complexities of what it means to live within the Jesus Tribe and in the shadow of the American Empire.

978-1-57312-592-5 208 pages/pb **$17.00**

Joint Venture
Jeanie Miley

Joint Venture is a memoir of the author's journey to find and express her inner, authentic self, not as an egotistical venture, but as a sacred responsibility and partnership with God. Miley's quest for Christian wholeness is a rich resource for other seekers.

978-1-57312-581-9 224 pages/pb **$17.00**

Let Me More of Their Beauty See
Reading Familiar Verses in Context
Diane G. Chen

Let Me More of Their Beauty See offers eight examples of how attention to the historical and literary settings can safeguard against taking a text out of context, bring out its transforming power in greater dimension, and help us apply Scripture appropriately in our daily lives.

978-1-57312-564-2 160 pages/pb **$17.00**

Looking Around for God
The Strangely Reverent Observations of an Unconventional Christian
James A. Autry

Looking Around for God, Autry's tenth book, is in many ways his most personal. In it he considers his unique life of faith and belief in God. Autry is a former Fortune 500 executive, author, poet, and consultant whose work has had a significant influence on leadership thinking.

978-157312-484-3 144 pages/pb **$16.00**

Maggie Lee for Good
Jinny and John Hinson

Maggie Lee for Good captures the essence of a young girl's boundless faith and spirit. Her parents' moving story of the accident that took her life will inspire readers who are facing loss, looking for evidence of God's sustaining grace, or searching for ways to make a meaningful difference in the lives of others.

978-1-57312-630-4 144 pages/pb **$15.00**

To order call **1-800-747-3016** or visit **www.helwys.com**

Mount and Mountain
Vol. 1: A Reverend and a Rabbi Talk About the Ten Commandments
Rami Shapiro and Michael Smith

Mount and Mountain represents the first half of an interfaith dialogue—a dialogue that neither preaches nor placates but challenges its participants to work both singly and together in the task of reinterpreting sacred texts. Mike and Rami discuss the nature of divinity, the power of faith, the beauty of myth and story, the necessity of doubt, the achievements, failings, and future of religion, and, above all, the struggle to live ethically and in harmony with the way of God. 978-1-57312-612-0 144 pages/pb **$15.00**

Overcoming Adolescence
Growing Beyond Childhood into Maturity
Marion D. Aldridge

In Overcoming Adolescence, Marion Aldridge poses questions for adults of all ages to consider. His challenge to readers is one he has personally worked to confront: to grow up *all the way*—mentally, physically, academically, socially, emotionally, and spiritually. The key involves not only knowing how to work through the process but also how to recognize what may be contributing to our perpetual adolescence.

978-1-57312-577-2 156 pages/pb **$17.00**

Psychic Pancakes & Communion Pizza
More Musings and Mutterings of a Church Misfit
Bert Montgomery

Psychic Pancakes & Communion Pizza is Bert Montgomery's highly anticipated follow-up to Elvis, Willie, Jesus & Me and contains further reflections on music, film, culture, life, and finding Jesus in the midst of it all. 978-1-57312-578-9 160 pages/pb **$16.00**

Reading Job (Reading the Old Testament series)
A Literary and Theological Commentary
James L. Crenshaw

At issue in the Book of Job is a question with which most all of us struggle at some point in life, "Why do bad things happen to good people?" James Crenshaw has devoted his life to studying the disturbing matter of theodicy—divine justice—that troubles many people of faith.

978-1-57312-574-1 192 pages/pb **$22.00**

To order call 1-800-747-3016 or visit www.helwys.com

Reading Samuel (Reading the Old Testament series)
A Literary and Theological Commentary
Johanna W. H. van Wijk-Bos

Interpreted masterfully by preeminent Old Testament scholar Johanna W. H. van Wijk-Bos, the story of Samuel touches on a vast array of subjects that make up the rich fabric of human life. The reader gains an inside look at leadership, royal intrigue, military campaigns, occult practices, and the significance of religious objects of veneration.

978-1-57312-607-6   272 pages/pb   **$22.00**

The Role of the Minister in a Dying Congregation
Lynwood B. Jenkins

In *The Role of the Minister in a Dying Congregation* Jenkins provides a courageous and responsible resource on one of the most critical issues in congregational life: how to help a congregation conclude its ministry life cycle with dignity and meaning.

978-1-57312-571-0   96 pages/pb   **$14.00**

Sessions with Philippians (Session Bible Studies series)
Finding Joy in Community
Bo Prosser

In this brief letter to the Philippians, Paul makes clear the centrality of his faith in Jesus Christ, his love for the Philippian church, and his joy in serving both Christ and their church.

978-1-57312-579-6   112 pages/pb   **$13.00**

Sessions with Samuel (Session Bible Studies series)
Stories from the Edge
Tony W. Cartledge

In these stories, Israel faces one crisis after another, a people constantly on the edge. Individuals such as Saul and David find themselves on the edge as well, facing troubles of leadership and personal struggle. Yet, each crisis becomes a gateway for learning that God is always present, that hope remains.

978-1-57312-555-0   112 pages/pb   **$13.00**

Silver Linings
My Life Before and After Challenger 7
June Scobee Rodgers

We know the public story of *Challenger 7*'s tragic destruction. That day, June's life took a new direction that ultimately led to the creation of the Challenger Center and to new life and new love. Her story of Christian faith and triumph over adversity will inspire readers of every age.

978-1-57312-570-3   352 pages/hc   **$28.00**

To order call 1-800-747-3016 or visit www.helwys.com

Spacious
Exploring Faith and Place
Holly Sprink

Exploring where we are and why that matters to God is an incredible, ongoing process. If we are present and attentive, God creatively and continuously widens our view of the world, whether we live in the Amazon or in our own hometown.

978-1-57312-649-6 156 pages/pb **$16.00**

This Is What a Preacher Looks Like
Sermons by Baptist Women in Ministry
Pamela Durso, ed.

In this collection of sermons by thirty-six Baptist women, their voices are soft and loud, prophetic and pastoral, humorous and sincere. They are African American, Asian, Latina, and Caucasian. They are sisters, wives, mothers, grandmothers, aunts, and friends.

978-1-57312-554-3 144 pages/pb **$18.00**

To Be a Good and Faithful Servant
The Life and Work of a Minister
Cecil Sherman

This book offers a window into how one pastor navigated the many daily challenges and opportunities of ministerial life and shares that wisdom with church leaders wherever they are in life—whether serving as lay leaders or as ministers just out of seminary, midway through a career, or seeking renewal after many years of service.

978-1-57312-559-8 208 pages/pb **$20.00**

Transformational Leadership
Leading with Integrity
Charles B. Bugg

"Transformational" leadership involves understanding and growing so that we can help create positive change in the world. This book encourages leaders to be willing to change if *they* want to help transform the world. They are honest about their personal strengths and weaknesses, and are not afraid of doing a fearless moral inventory of themselves.

978-1-57312-558-1 112 pages/pb **$14.00**

Written on My Heart
Daily Devotions for Your Journey through the Bible
Ann H. Smith

Smith takes readers on a fresh and exciting journey of daily readings of the Bible that will change, surprise, and renew you.

978-1-57312-549-9 288 pages/pb **$18.00**

To order call 1-800-747-3016 or visit www.helwys.com

When Crisis Comes Home
Revised and Expanded
John Lepper

The Bible is full of examples of how God's people, with homes grounded in the faith, faced crisis after crisis. These biblical personalities and families were not hopeless in the face of catastrophe—instead, their faith in God buoyed them, giving them hope for the future and strength to cope in the present. John Lepper will help you and your family prepare for, deal with, and learn from crises in your home. 978-1-57312-539-0 152 pages/pb **$17.00**

Cecil Sherman Formations Commentary

Add the wit and wisdom of Cecil Sherman to your library. He wrote the Smyth & Helwys Formations Commentary for 15 years; now you can purchase the 5-volume compilation covering the best of Cecil Sherman from Genesis to Revelation.

Vol. 1: Genesis–Job   1-57312-476-1 208 pages/pb **$17.00**
Vol. 2: Psalms–Malachi   1-57312-477-X 208 pages/pb **$17.00**
Vol. 3: Matthew–Mark   1-57312-478-8 208 pages/pb **$17.00**
Vol. 4: Luke–Acts   1-57312-479-6 208 pages/pb **$17.00**
Vol. 5: Romans–Revelation 1-57312-480-X 208 pages/pb **$17.00**

To order call **1-800-747-3016** or visit **www.helwys.com**

Clarence Jordan's Cotton Patch Gospel

The Complete Collection

Hardback • 448 pages
Retail ~~50.00~~ • Your Price 45.00

The Cotton Patch Gospel, by Koinonia Farm founder Clarence Jordan, recasts the stories of Jesus and the letters of the New Testament into the language and culture of the mid-twentieth-century South. Born out of the civil rights struggle, these now-classic translations of much of the New Testament bring the far-away places of Scripture closer to home: Gainesville, Selma, Birmingham, Atlanta, Washington D.C.

More than a translation, *The Cotton Patch Gospel* continues to make clear the startling relevance of Scripture for today. Now for the first time collected in a single, hardcover volume, this edition comes complete with a new Introduction by President Jimmy Carter, a Foreword by Will D. Campbell, and an Afterword by Tony Campolo. Smyth & Helwys Publishing is proud to help reintroduce these seminal works of Clarence Jordan to a new generation of believers, in an edition that can be passed down to generations still to come.

SMYTH & HELWYS

To order call **1-800-747-3016**
or visit **www.helwys.com**

www.ingramcontent.com/pod-product-compliance
Lightning Source LLC
Chambersburg PA
CBHW071723090426
42738CB00009B/1861